SMALL CHANGE

by

Marianne Frances

illustrated by Lesley Skeates

Green Print
London

Published February 1990
The Centre for Human Ecology, Edinburgh

Copyright © Marianne Frances and Lesley Skeates

This edition first published in 1993 by
Green Print
The Merlin Press Ltd
10 Malden Road
London NW5 3HR

ISBN 185425 081 7

Typeset by Computerset, Harmondsworth, Middlesex
Printed by Biddles Ltd., Guildford, Surrey

CONTENTS

INTRODUCTION

'It is better for many to move forward a few inches than for a few to move a mile'

Ask anyone nowadays and you can expect them to agree that, as individuals, we all share responsibility for the environment. We realise that it is no longer enough only to be aware: we must turn awareness into action. This is a new feeling – a feeling of wanting to help actively, of wanting to be involved in creating a better and fairer world, of readiness to take personal responsibility for the environment in everyday life, in short, a will to DO SOMETHING. Many people, including many who would not think of themselves as particularly 'green', are asking WHAT CAN I DO TO HELP? – a question that could be the most important one to be asked in this last decade of a thousand years. This book attempts some small answers.

Until recently, many people used to think that what individuals can do is too little to be of any consequence; so why bother? Why bother to save bits of foil when industry uses up millions of tons of aluminium? Why go out of your way to buy phosphate-free detergent when annually thousands of tons of phosphates are spread over fields? Why try to cut your mileage when road transport is being expanded year by year? But sceptical attitudes such as these are changing and every action, however small, towards environmental conservation is now seen as a worthwhile contribution. Worthwhile for many reasons – because small actions by millions do add up to something big and because small actions prepare us for the bigger changes that we will have to make before long, if all people are to have their rightful share in life and if humankind is not to destroy the masterpiece of the living world. Although conservation on a global scale must become every government's highest priority, it is important that every one of us participates at the level of everyday life. We can even lead the way.

This book takes a tour round the home. Each room gives an opportunity to look at different environmental problems and to suggest ways in which we can help. The tour then ventures outside to consider other activities. Before going outside we stop for question time, to consider why we are bothering and what sort of a lifestyle 'small change' might lead to.

As the title suggests, the actions described in this book are small-scale so that they can fit immediately and smoothly into the busy and non-technical individual's everyday life. Therefore there are no suggestions for actions which need money and expertise, such as installing solar panels, or actions that are as yet unrealistic for most of us such as giving up travelling by car, even though this is what we should be working towards. Similarly there is no mention of letter writing to MP's or PM's, undeniably an invaluable activity, but one that most of us find difficult to get round to doing.

What you will find in SMALL CHANGE are samples of actions that can be done easily. A short book must be selective; it cannot consider every environmental problem nor every part of the home, let alone all of the hundreds of products we use. It is hoped that it will inspire readers to think of many further actions for themselves, for one thing is clear – if we are to limit our use of resources, we must develop unlimited resourcefulness! Not all the suggestions will be appropriate

1

for everyone, nor will all appeal to everyone – just like with a cookery book, you can pick and choose and modify the recipes. Although many of the suggestions are about what to buy and what to avoid buying, there is lots of scope for actions which have little or nothing to do with spending. Ways of helping the environment are open to everyone.

A sustainable environment, the conservation of nature's heritage of plants and animals, as well as justice to the people of the developing world, are the themes underlying this book. Some issues however, such as world population, armaments and nuclear power, being difficult to act on at everyday level, are omitted even though they are vital to our survival. Nor do the suggestions for actions aim directly at improving health or creating 'better living'; not because these are held to be unimportant, but because it is more and more apparent to me that in all aspects of human activity, the best interests, health and fulfilment of humankind and that of the planet happily coincide.

Marianne Frances, January 1993

ACKNOWLEDGEMENTS

Many people encouraged and helped me with this book. Among them are personal friends in Scotland and in the US; environmental friends in Friends of the Earth (Scotland), in Turning Point, in The Future in our Hands, in The Centre for Human Ecology of Edinburgh University. Linda Muir's criticisms livened up the text; Lesley Skeates' irresistible drawings make the book. My warmest thanks to all.

the kitchen

THE KITCHEN

'What's that? Me and my family change the world? People like us? By making . . . small changes? I've already got enough to do, what with work, shopping, cooking, doing up the house, driving the children here, there and everywhere, birthdays, Christmas. . . .' But everything we do has an effect somewhere in the world; it is up to us whether it will be for better or for worse.

Shopping for food

Every purchase we make, or avoid making, for food or any other items, sends out information that spreads like a ripple – first to retailers about what and what not to stock, then to producers indicating which products we find to our liking, and then on to powerful companies the world over whose economic activities shape the environment and control people's lives. When shopping for food we can make choices that will help to restore the environment, give people everywhere a better deal, promote animal welfare, while at the same time be healthier for ourselves. The 'environmentally friendly' shopper needs not only to be cost, quality and health conscious but needs in addition to keep the whole planet in mind! Asking a lot? But we do ask a lot – of the planet.

But there's yet another point to consider. It's not only WHAT we buy but also WHERE we buy that has far reaching effects on the environment. Our preference to shop at big centres (supermarkets, malls, out of town centres) or in our neighbourhood shops influences city planning, sets the pattern of public and private transport, affects local employment, local agriculture, all of which helps or hinders environmental conservation.

Supermarket or Local Shops?

From the shopper's point of view there are obvious advantages in shopping for food in streamlined supermarkets. The price of food is generally lower than in small shops because of bulk purchasing and big scale operation. Products are standardised and reliably available. It is indeed very convenient to pick up all you need under one roof. But this convenience is most available to car owners. Supermarket shopping encourages car ownership and supermarket car parks and access roads bury acres of valuable land under asphalt.

There are many environmental and social costs of centralised shopping. Money spent in supermarkets and chain stores drains away to remote companies and producers and brings little direct benefit to the local community. Many traditional

grocers and butchers are forced out of business. Supermarkets employ relatively few people locally (after completion of the buildings) and mainly deal with large suppliers, however distant, rather than smaller local ones. The very features which attract shoppers to supermarkets – low prices, convenience, reliable supply and standardisation of products – are contributing to environmental changes for the worse in country and city. Although supermarket shopping is part of our way of life, and some now sell organically grown foods, small wholefood shops are nevertheless very popular. Could this show that a change in shopping habits is possible? The many clothes boutiques that opened in the 60's and 70's contributed to the closing of a number of traditional department stores. If it can happen with clothes, might it not happen with food?

WHAT YOU CAN DO

Going Shopping

- *If you are going shopping by car, could you go with a friend? Car sharing makes sense.*

- *Avoid making frequent car journeys to supermarkets for small amounts of shopping.*

- *Support local shops – they are convenient and made your neighbourhood a lively community. Keep fit by walking or cycling to local shops.*

- *Before buying tea and coffee read page 8! You can buy a number of foods and household goods, imported from the developing world through 'fair trade' agreements, at OXFAM shops (or various charity shops). Every such purchase you make supports people and the environment. Or order by catalogue (for addresses, see page 80).*

- *If you have managed to save time and money by shopping at a supermarket, use some of both for an environmental purpose! Write for a 'Fairtrade' catalogue, find out about recycling schemes and other local environmental activities. . . .*

Buying Food – In Big or In Small Quantities?

- *Big quantities use less packaging materials, weight for weight, but you need storage space and facilities (which may use up energy, see 'Storing Food' page 18). However it is tempting to use more because it's there and bulk buying usually means supermarket shopping. Buy in reasonably large quantities.*

- *Small quantities use more packaging, weight for weight. Small individually wrapped items are most wasteful of packaging and are responsible for a lot of litter. Avoid individually wrapped items, sachets, etc.*

Buying Food – Loose or Packaged

- *Few things other than fruit and vegetables can be bought loose nowadays. But always take some polybags with you to buy unpackaged goods where possible.*

Buying Food – In Glass, Metal, Paper or Plastic?

- *The final destination of packaging is (nearly) as important to think about as the destination of the contents! Is it*
  > Recyclable?
  > Re-usable?
  > Burnable?
  > Bio-degradable?
  Or will the packaging finish up on the rubbish mountain?

FORGET~ME~NOTS!

You are conserving resources and being considerate of the environment if you

- *Buy drinks in RETURNABLE bottles and RETURN them.*

- *Take non-returnable empties to recycling 'banks' (glass and metal banks are now provided in many places, plastic and fabric collection is beginning).*

- *Choose items wrapped simply in paper or cardboard because these can be burnt or will rot away.*

- *Avoid plastic if there is a choice of other container (milk, for example).*

- *Avoid polystyrene foam packaging (trays for fresh foods, containers for fast food). Even though most polystyrene is now made using a gas that is (only somewhat) less harmful to the ozone layer, it is a wasteful form of packaging. Surprise your local carry-out shop by coming with last time's washed foil trays or your own dishes!*

- *Consider carefully the need to use disposables.*

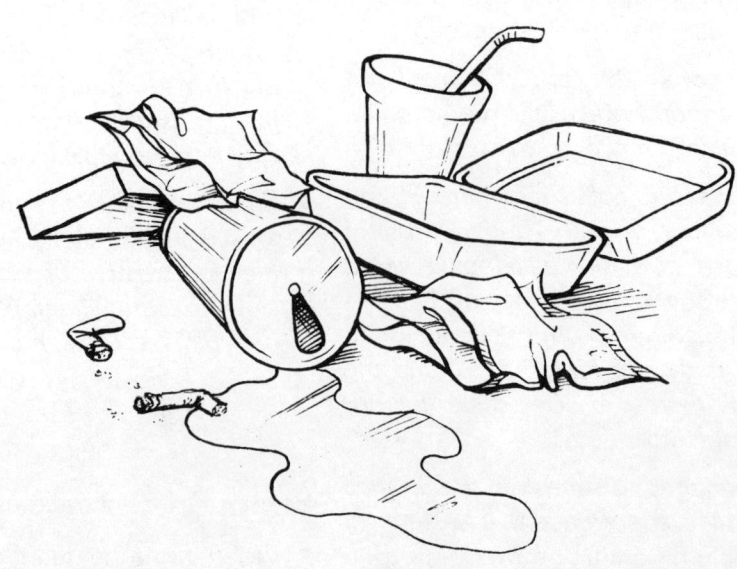

Food – Your Choice Can Help

Good food is the foundation of good health and it is necessary that we make the right choices for our general well-being. This is common sense. It may not be so obvious that the well-being of the environment also depends on making those same choices. The sort of food and drink we consume at present and the way we grow it and process it, is damaging land and sea more than any other human activity. In choosing food grown in a way that helps to conserve the environment we are also choosing food which is healthier for us.

In the same way, changes in our eating and drinking habits would make more food available to people in the developing world. We depend greatly on food imports from these countries: tea, coffee, cane sugar, bananas and other tropical fruits, peanuts and other nuts, rice . . . and feedstuffs for our animals. These are grown as cash crops for export, usually taking good land which should instead be growing food for the country's own needs. Many of the world's severest environmental problems such as soil erosion, deforestation, the spreading of deserts, pollution of water, with all the terrible consequences for people and wildlife, are the direct (and indirect) result of growing cash crops for the countries of the affluent world. By choosing what we eat and drink with consideration, we can change things for the better in both our and the developing world.

It would take volumes to go into the environmental background of the great variety of foods and drinks that we ordinarily consume. Instead, join a tour through a day's meals which looks at how a selection of the foods you may be eating affects the environment. What small changes could we make?

BREAKFAST
Do You Start Breakfast With Fruit Juice? Orange, Grapefruit or Apple?

Until recently a glass of orange or grapefruit juice, squeezed by hand, was something special, reserved for Sunday breakfast or a meal out. Now many people drink fruit juice freely, not only at meals but whenever they feel thirsty. Citrus juice production has expanded hugely in recent years.

In Central America, tropical forests have been cleared to make way for citrus plantations and around the Mediterranean hillsides have been stripped of their natural cork oak forest to grow oranges for juice. A pleasant and healthy drink adds its bit to global deforestation. Many apple varieties, on the other hand, grow splendidly in Britain and apple trees enhance the environment, so. . . .

7

WHAT YOU CAN DO

Encourage British apple growers to restore the famous British apple orchards and the world famous varieties of British apple.

Discourage the multinational corporations that produce and market citrus fruit drinks from destroying tropical forests.

- Drink British apple juice, made from organically grown apples, if obtainable, in preference to citrus juice.

- Reserve citrus juice for occasional use; eat oranges or grapefruit (and apples) whole instead of drinking only the juice and have the benefit of roughage as well.

Do You Take Tea or Coffee?

Tea from Darjeeling, Poonakandy, Sri-Lanka, China: coffee from Kenya, Brazil, Columbia – the names on the packets conjure up far-off romantic places!

The reality of tea and coffee growing is far from romantic. The production process, starting at the tea estates and coffee growers in the developing world and ending at shops in Europe, America, Australia and New Zealand is wholly controlled by a few large (transnational) corporations. Their goal is to make profits while selling tea and coffee to us at relatively low prices*. The losers are the tea estate workers and small coffee growers who work for minimal wages and live in impoverished conditions. The spread of tea plantations is driving more small farmers off the land into destitution.

There is an alternative to buying from the giant companies. Traidcraft, Oxfam, Twin Trading and Equal Exchange import coffee and tea on 'Fairtrade' terms from small growers, or from estates which offer their workers reasonable conditions. Fairtrade tea and coffee is sold by the organisations mentioned (for list, see page 80) and quite a few wholefood shops. One brand, Cafe Direct, is now on sale at Safeways.

WHAT YOU CAN DO

- Use tea and coffee carefully; don't make a lot more than you will drink.

- Buy Fairtrade coffee and tea an buy some to introduce to a friend (see page 80 for addresses of Fairtrade organisations).

- If you buy commercial brands, buy good quality grades. It is the cheap grades that exploit the most (this is true of most goods).

- Use fresh coffee rather than instant (see Midmorning Break page 10 for reason why).

- Have you tried alternative drinks? Barleycup, Caro, Yannoh instead of coffee: herb teas instead of black tea.

---

*Yes indeed! Tea and coffee ARE cheap. How often do we pour away an unfinished potful of tea or leave half a mug of coffee standing – we wouldn't do that if the price was high. Pour an unfinished bottle of whisky down the drain? Leave half the drachm in your glass? Not often, I bet!

Breakfast Cereals or Muesli?

The cereals, which tumble into millions of British breakfast bowls every, morning have run up a very big energy bill in production. Breakfast cereals are made mainly out of processed wheat (or maize) and sugar and have copious additions of synthetic vitamins. Muesli is a mixture of rolled wheat, oats and rye with added dried fruit and nuts. Many brands are sugar-free and most are free of added vitamins. Far less energy is needed to produce muesli because it is minimally processed and additive-free, still less if the ingredients have been grown 'organically'. It's also healthier for teeth than processed cereals. Environmentally, muesli has a lot to recommend it.

WHAT YOU CAN DO

- *If you are used to processed cereals, muesli may take a little getting used to. It's worth trying, and with muesli for breakfast you may not need a mid-morning snack.*

- *Make up your own muesli mixture. Ingredients are obtainable from wholefood shops. The most basic form of muesli is a mixture of uncooked rolled oats with grated apple.*

- *Try cereals or muesli with apple juice (British organic) instead of milk (intensively produced, see Lunch page 11).*

An Egg For Breakfast

Each person in Britain eats an average of 250 eggs per year (many 'concealed' in home cooked and commercially prepared food). 96% of all eggs are produced in battery 'farms'. Hens raised in this intensive way need to be fed on grain (which could feed many more people directly as bread or cereal) and fishmeal (which depletes the sea of fish, see SUPPER, page 13). Battery farming of poultry was the first 'agribusiness' to make the public aware of inhumane mass-production of animals. Producers say that the public demand for cheap eggs justifies battery farming but competition between egg producers is equally to blame. The Salmonella epidemic of 1988 (and recently 'mad cow' disease amongst cattle) has made us question cost cutting practices in intensive animal raising.

WHAT YOU CAN DO

- *Use fewer eggs. Apart from health risks, it would be better for the environment and hens themselves if we ate fewer eggs.*

- *Buy free-range eggs. Free-range is not necessarily 'organic' and conditions on some F-R farms are far from good but it is a step in the right direction.*

MID-MORNING BREAK

A quick cup of instant coffee? A whole extra industrial process is needed to convert ground into instant coffee. Compare the 'energy bills'.

	INSTANT COFFEE	GROUND COFFEE
Energy for roasting and grinding beans	X	X
Energy for preparing coffee extract	X	—
Energy for evaporating extract to a powder	X	—
Energy to manufacture glass jars for packaging	X	—
ENERGY BILL	XXXX	X

After this big expenditure of energy to manufacture instant coffee, we boil water yet again to redissolve it and then – wait for it to cool!

WHAT YOU CAN DO

- *Use instant coffee sparingly.*
- *Boil only as much water as is needed.*

- *Return instant coffee jars to a bottle bank or refill from refill packets.*

FAST SNACKS
QUICKER LITTER

LITTER

Have a Snack

In Britain we munch our way through £1 billion of savoury snacks each year. We each eat an average of 100 packets of crisps per year and a small fortune's worth of snacks. Most snacks are highly processed and packaged foods which need a lot of energy to manufacture and £millions to advertise. The ones that most appeal to children are often the ones with most additives ('junk foods'). Most of the litter blowing around city streets and school playgrounds is discarded snack packaging.

WHAT YOU CAN DO

- *Take a sandwich (re-using a plastic bag!) or nuts and raisins or other dried fruit instead.*
- *Have an apple or other fruit already conveniently packaged by Nature.*
- *By eating Brazil nuts you help to conserve tropical forests! Brazil nuts grow only in intact tropical forest, so trade in nuts creates an economic reason for conservation – a small one perhaps, but every little helps.*

10

LUNCH
A Meat Dish or Beans on Toast?

Our high regard for meat (and animal products) has done more to shape the countryside than the growing of any other food. Most of England, lowland Scotland and Wales is a shore-to-shore patchwork of fields. Most of these fields serve animal production, either as pasture or by growing fodder crops; only one field in ten produces food for people directly (some grains, potatoes, beans and vegetables). In addition we import millions of tons of feedstuffs, much of it from developing countries. Meat production dominates agriculture worldwide: everywhere domestic animals compete with forests and wildlife and the basic needs of people. PRODUCING and CONSUMING LESS meat, less dairy products and fewer eggs would release land in the developing countries and enable people there to feed themselves better. In Britain we could grow more hardwood trees for badly needed timber and eventually reduce our enormous imports (see LIVING ROOM page 37). Our countryside could support more wildlife and provide more opportunities for recreation. Raising fewer animals would assure them better treatment since many inhumane practices are the result of pressures of mass production. One person in three in Britain already chooses to eat less red meat for personal and environmental reasons, though consumption of poultry (white meat) has gone up. For the better health of the planet and for improvement in the standard of world nutrition, it will be necessary for most of us to reduce our consumption of ALL meat and ALL animal products and to obtain more protein from plants. Long live beans!

WHAT YOU CAN DO

- *Make meat go further. Stews and Italian pasta dishes make good use of small amounts of meat.*

- *Replace at least one meat meal each week with fish (see SUPPER page 13 for reason why).*

- *Discover bean cookery. Beans are the best replacement for animal protein. Get a bean cookery book.*

- *For a meal out choose a Chinese or other Far Eastern restaurant whose cooking uses meat and fish sparingly.*

- *Change the idea that a Square Meal must be 'Meat and two Veg.'. 'Two Veg. with some Meat' can be just as sustaining if the dish contains beans, brown rice, barley, nuts or seeds.*

AFTERNOON BREAK
Time for a smoke? or a sweetie!

Tobacco plants are so hungry for nutrients and water that a crop quickly drains the soil of fertility. The plants are very prone to disease and so must be sprayed frequently (putting cultivators' life and health at risk from pesticides). Growing tobacco takes up land and resources which could be used to grow food; smoke-curing tobacco needs huge amounts of wood. Around one million hectares of forest are cut each year to provide wood for curing. It is clear that tobacco has already done a lot of damage long before smoking it damages people's health. This is a planetary environmental health warning!

WHAT YOU CAN DO

- *Think about the serious environmental effects of tobacco production and manufacture next time you light up.*

- *Do all in your power to prevent a young person from starting to smoke.*

- *If you do smoke, please smoke well away from others particularly children and non-smokers (and especially smokers who have just succeeded to give up!).*

So What About a Sweetie Instead?

Most people, and many animals, have a sweet tooth! Honey has always been highly prized but only in recent times has sugar become thought of as a regular, basic foodstuff. In Britain an adult eats nearly his/her own weight in sugar each year (100lbs. per person per year). Many children are probably eating more than their own weight each year! However, sugar consumption in the affluent world in general is falling.

Sugar cane is a cash crop and growing cash crops on a huge scale in tropical countries creates many ecological and economic problems. Nevertheless sugar cane is important to the economy of the humid tropics because it is a resilient plant which stands up to hurricanes and tropical storms better than most other crops. Sugar beet grown in Britain and Europe has caused the steady decline of trade of many sugar cane growing countries who are now experiencing great poverty as a result. As we lower our sugar consumption, we should see that what we do still buy is cane sugar, so that the sugar cane countries have time to develop alternative trade.

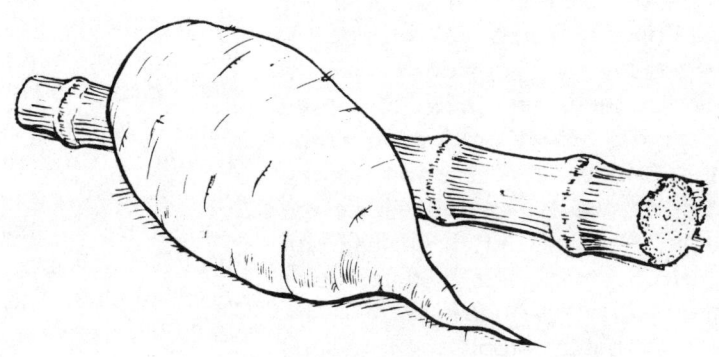

WHAT YOU CAN DO

- *Eat nuts and dried fruit instead of sweets. 'Fairtrade' chocolate is now available (see page 8 and page 80), also a very strong organic chocolate (from wholefood shops).*

- *Buy cane sugar where there is a choice; it says which it is on the packet.*

- *Use half the quantity given in (most) recipes for cakes, pastry, biscuits, home-made drinks.*

- *Use honey to replace some sugar. Support the environmentally useful honey bee!*

- *Make lemonade from lemons instead of buying sugar-rich and expensive soft drinks. Use a little (cane) sugar if necessary. This way you avoid chemical sweeteners and additives and you aren't adding more plastic bottles to the rubbish mountain. Also, it's cheaper.*

- *Reserve icing (cane sugar) cakes for special occasions. Use Apple and Pear Spread (available in wholefood shops) in place of icing sugar and sprinkle on grated or chopped nuts or desiccated coconut.*

SUPPER
Fish and Chips: eat 'e'fish'ntly'!

Fish is good for you and it is good for the conservation of life in the ocean to eat fresh fish instead of meat. A paradox? How can eating fish conserve fish? What has fish to do with meat?

Almost half of the world's entire seafish catch is processed into fishmeal and oil for feeding to intensively raised livestock (pigs and poultry and, surprisingly, farmed fish). As much as two thirds of the North Sea fish catch ends up feeding animals (including cats). If people ate more fish direct, instead of first converting fish (ine'fish'ntly!) into animals, far fewer would need be caught. Fish stocks could regain their proper levels and the vast, but not limitless, resources of the seas could be sensibly utilised, to the lasting advantage of fishermen, consumers and marine life.

Fish farming, if managed with proper concern for the environment, could greatly help marine conservation. It is very wasteful to trawl ever greater distances at sea, using up huge quantities of fuel and fishing by methods that cause serious ecological damage. Factory ships ranging the oceans can easily elude environmental control and continue cruel practices such as drift netting (see Bedroom page 30). But at present fish farms are creating rather than solving environmental problems. They pollute water with excreta and chemicals; animals and birds suspected of taking fish are killed. Hopefully, informed public pressure will persuade this industry, which could be valuable to the environment, to operate more ecologically.

WHAT YOU CAN DO

- *Eat fish regularly in place of meat or eggs at some meals each week. Buy sea fish, rather than farmed fish, at present.*

- *Try some of the less usual varieties. For lack of a market, these tasty fish are often thrown back into the sea (already dead). Eating these would reduce the amount of the popular white fish varieties that need to be caught.*

- *Buy 'line' caught fish in preference to netted fish. Catching by line does much less environmental damage and conserves fish stocks. Ask your fishmonger which fish is which.*

- *Make fish go further in dishes such as kedgeree, fish soup, fish cakes, Chinese dishes.*

- *Avoid eating tuna until drift netting is totally banned in all countries. 'Dolphin-friendly' tuna may not be altogether free of drift-netted product.*

And Chips

The modern potato is a perfectly naturalised British vegetable that would hardly recognise its distant Peruvian relatives. A whole potato is a wholefood, containing carbohydrates, protein, vitamins and minerals; it is designed by nature for winter storage. But caution! Stored potatoes are nowadays treated with fungicide and should be peeled, so the most nutritious part of the potato has to be discarded. For this reason it is worth growing your own. There are some interesting varieties to grow, such as –

Fish and chips as a food scores environmental good marks, and would score better health marks too if vegetable oils high in polyunsaturates (well drained) are used for frying.

13

WHAT YOU CAN DO

- *Find a source of organically grown potatoes. Because we eat a lot of them, it is more important that potatoes are organically grown than other vegetables of which we may eat only a little.*

- *Try to buy some of the less well-known potato varieties and encourage potato growers to diversify their crops (an ecologically good practice).*

- *Avoid using powdered potato (for the same reason as avoiding instant coffee, see Mid-morning break, page 10).*

- *Grow some of your own potatoes.*

What Does This All Add Up To?

This brief tour of the day's meals has selected some examples to show how our food affects both the environment here and worldwide and the wellbeing of the countries we depend on. Here are some guidelines:

- *It isn't necessary to stop eating meat altogether but it is necessary for a great many of us to eat less meat (of all types), less dairy products and fewer eggs. Eat fresh fish in place of some meat and eggs. Use protein rich beans and lentils in place of meat.*

- *Eat more vegetables and salads.*

- *Buy as much organically grown food as you can get (or afford).*

- *Eat more fruit and nuts in place of milky, eggy rich puddings.*

- *If possible buy what is locally in season for everyday meals. Drink British apple juice.*

- *Reserve more 'exotic' varieties of fruit and vegetables, which have been flown in from tropical countries, for special occasions.*

- *Use foods imported from developing countries thoughtfully and sparingly. Don't waste tea and coffee, drink less citrus juice, use less sugar.*

- *Buy 'Fairtrade' food items in place of the same items imported by multinational companies from the developing world.*

Cooking: What Energy?

Cookers and kettles are some of the most frequently used appliances in the home. Although cooking consumes a lot less energy than space heating or water heating, cooking makes us directly conscious of fuel use and energy saving. It is something we can watch happening and can control and something we do every day of the year. Moreover, the more you avoid convenience foods, which need minimal cooking and the more environmentally friendly your meals, the more your cooker will be in action!

Gas (North Sea) or Electricity?

Around 55% of households cook with gas, about 41% with electricity though some of these households might choose gas if it was available in their area. Most people prefer cooking with gas, despite big improvements in speed and efficiency of electric hotplates. Cooking with electricity can be made very energy efficient in the kitchen because electric plates can exactly match pans, heating elements can be built into pans, casseroles and kettles and electric ovens can be well sealed and insulated. Gas appears to be more wasteful because gas burners inevitably spread the heat and gas ovens must be ventilated with air intake and outlet.

However when the environmental costs of producing and bringing electricity and gas to your home is taken into account, things look different.

Producing Gas

All gas used in Britain now comes from reserves of natural gas under the North Sea. It is an almost 'ready to use' fuel (technically known as a fuel with 'low energy overheads'). At present some gas is lost during distribution but this loss will become negligible when all old pipe lines are replaced. Gas burns cleanly to carbon dioxide and water. Carbon dioxide is of course one of the 'greenhouse' gases responsible for global warming, but the quantity produced by domestic use of natural gas is small compared to the quantities released from burning coal or oil at home or for electricity generation. It is as nothing compared to the carbon dioxide released by motor transport or burning the rainforests.

Producing Electricity

Generating electricity by burning fossil fuels makes inefficient use of their stored energy and releases quantities of polluting gases into the atmosphere. Nuclear generation of electricity is no answer to this problem because it is imposing a legacy of radioactive waste on untold future generations of life on earth.

'Best Choice for the Future'?

Electricity is a uniquely high grade energy and is environmentally costly to produce; it makes sense to use it only where it is indispensable or offers real advantages over other fuels. In the home small electric motors which drive refrigerators,

This picture shows that, if the energy lost in generating and distributing are reckoned in, cooking by electricity is not much more efficient than cooking on a camp fire.

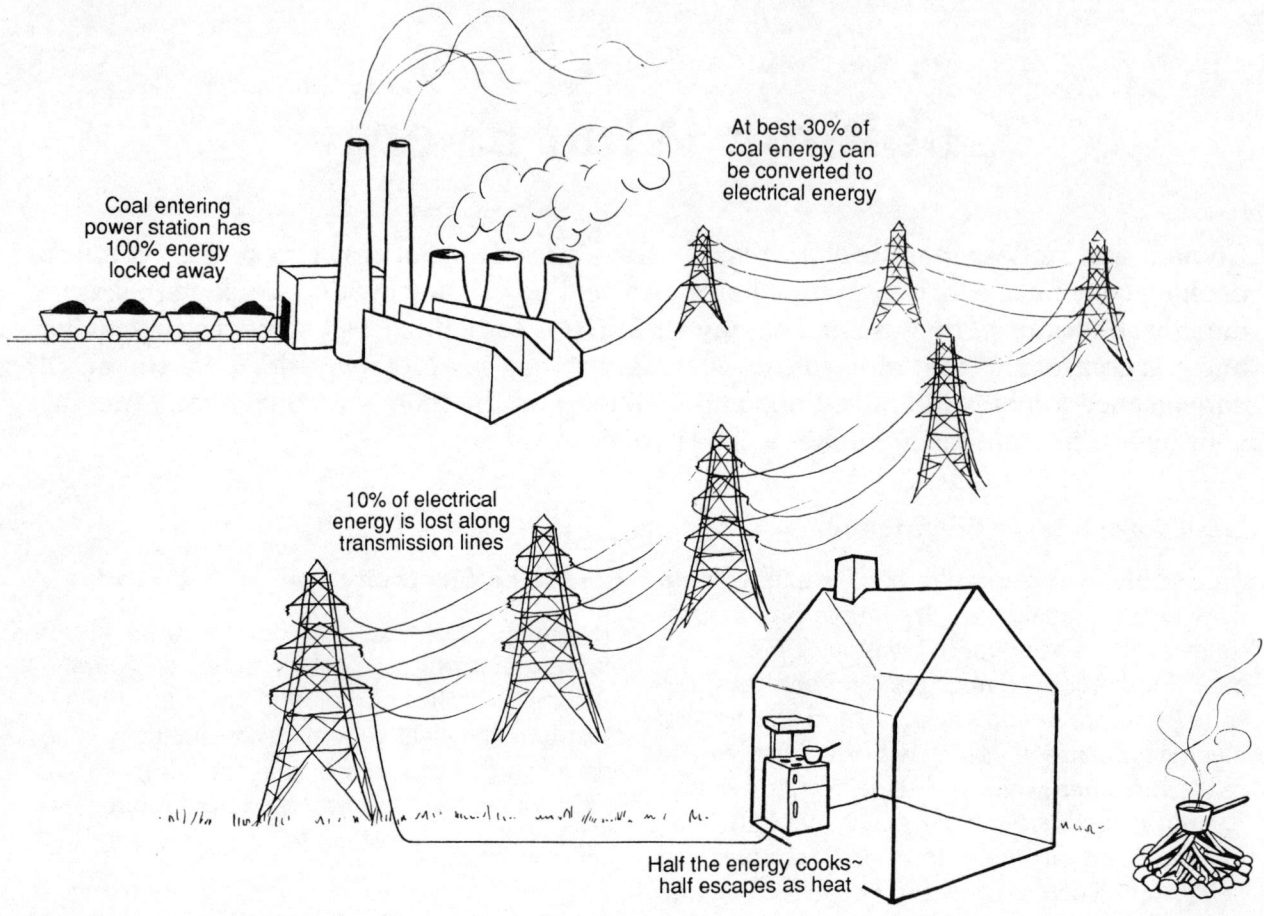

Coal entering power station has 100% energy locked away

At best 30% of coal energy can be converted to electrical energy

10% of electrical energy is lost along transmission lines

Half the energy cooks~ half escapes as heat

washing machinery, vacuum cleaners, food mixers, are a good use. Electric lighting is convenient, although ordinary, incandescent bulbs are not very efficient since they produce more heat energy than light energy (for low energy bulbs see Hall, page 33). Where continuous moderate (low grade) heat is needed, as for space and water heating or for cooking, the best choice for the environment is gas.

WHAT YOU CAN DO

There is a lot of scope for economising energy in cooking, whatever the fuel your cooker uses.

Trap the Heat

- *Always use a lid (or a plate) on pots and pans and turn down the heat.*

- *Don't put on a full kettle for one 'cuppa'.*

- *Use a pressure cooker as it cuts cooking time and has the added advantage of causing less damp as the steam is contained in the pot.*

- *Use minimum amounts of water to cook vegetables and thereby conserve both vitamins and fuel energy. Steamers enable you to cook several things on one burner/hotplate.*

- *Eat more raw or very lightly cooked vegetables. Save fuel and improve health.*

- *Kettles*, especially the jug sort, electric pans, casseroles, with built-in elements use electricity efficiently. 'Slow cookers' use no more electricity than a lightbulb and won't burn food, even if you are delayed getting home.*

---

*Electric kettles are quick and efficient but take a very large amount of power. If, as regularly happens, everyone switches on their kettle at about the same time, at breakfast, lunch and tea or at the end of a popular telly programme, there is a sudden big peak in demand for power. Such peaks influence the estimates of required generating capacity. It would make a difference to power planning if everyone boiled water by gas!

- *Avoid heating the oven for one small dish. If you need to do so, is there anything that could go in as well – bread crusts to dry (for crumbs), nuts to roast? Is pre-heating the oven necessary? Turn off an electric oven before the end of the cooking time.*

- *Try baking with a continental style cake tin with a well. This brings heat into the centre of the cake and so cuts cooking time and avoids soggy middles!*

What About Microwave Ovens?

Microwave ovens use much the same power as ordinary electric ovens but since food cooks much faster, there is a considerable saving of energy. In addition there will be fewer encrusted pans and dishes to wash up, so saving on hot water and detergents. This much is in favour of the microwave oven as an environmentally friendly invention. But consider the following:

- *They are essentially additional appliances for which you may need to buy special dishes. Most people will still use their electric or gas cookers. Manufacture of any appliance has its environmental costs in resources and energy and every manufacturing process causes a bit more global pollution.*

- *Microwave ovens encourage you to buy convenience foods, especially frozen, ready cooked meals. This in turn encourages more use of freezers. Convenience foods make for a lot of packaging rubbish.*

- *One of the selling points for microwave ovens is that you can thaw out frozen foods in minutes. Of course doing this can be a 'life-saver' on occasions, but is an absurdly anti-energy-conservation practice if done regularly!*

Storing Food

People have always spent a lot of time and trouble storing food for lean times. In the past this meant winter, war or crop failure. Today we store food mainly for convenience, for economy or to give us variety independent of the seasons.

Modern ways of storing perishable food rely heavily on energy and resources. Some ways are more environmentally costly than others and we should consider each method and decide when it is worth using it. For example, is it really justifiable to wrap a half litre of liquid (soups, drinks, etc.) in metal or glass when the wrapping is often more valuable than the contents? Refrigeration uses nearly one fifth of all the power produced in the UK. Should we be using up so much fossil fuel and putting huge quantities of carbon dioxide into the atmosphere, to have the luxury of eating raspberries in February? We could save a lot of energy by eating with the seasons, and rediscover the pleasure of foods at their own special time of the year.

WHAT YOU CAN DO

- *For every day purposes, eat fresh food, locally grown to avoid transport and refrigerated storage.*

- *Keep tins only for emergencies.*

- *Buy off-season food that is intended by nature to be stored – potatoes, apples, nuts, grains, dried peas, beans and lentils, rice, etc.*

- *Reconsider whether to use a freezer.*

Fridges and Freezers

Follow the principle that the lower your electricity bill is, and the nation's electricity bill, the better for the environment (until electricity is generated from wind and wave power – and probably even then!).

- *Follow the manufacturer's instructions for maintenance and defrosting so that your fridge/freezer runs as efficiently as possible.*

- *Make sure that both fridge and freezer doors/lids close tightly. A thin smear of vaseline over door and lid seals will make them grip tightly.*

- *To reduce running costs, stand the fridge away from any source of heat, preferably in a cold part of the kitchen/house.*

- *If possible, put your freezer out of sight (in garage or garden shed); not so convenient perhaps, but it won't be opened so often!*

- *For normal running, set the fridge thermostat to minimum cooling – always allow warm food to cool before refrigerating or freezing.*

- *Plan ahead and put deep frozen food on the top shelf of the fridge to thaw slowly. This way the 'cold energy' produced by the freezer, or the fridge's own freezing compartment, contributes to the cold in the fridge instead of cooling down the kitchen.*

- *Could you turn off the fridge during the cold winter months, if your kitchen has a well ventilated larder? It is cheaper to throw out a few things that may go bad than to keep a fridge running over the winter months. Consider this seriously if you run a separate freezer. It may not be advisable to do this if you use a lot of pre-cooked and processed foods, especially processed meat and fish (pastes, pates and pies).*

Buying a New Fridge or Freezer

The most efficient fridges now use less than 150 units of electricity a year (sparing the atmosphere the release of a roomful of CO_2); the most efficient freezers use less than 350 units a year (sparing the atmosphere the release of 3 roomfuls of CO_2). Ask about the energy efficiency of the model you choose. Don't take 'it depends on how you use it' as a satisfactory answer! While this may be so, the basic energy consumption for running should be stamped on the label of the appliance. Energy efficiency is an important aspect to consider before buying any new electrical appliance – kettles, cookers, washing machines, driers, television sets, etc., but it is especially important for fridges and freezers, which are permanently running and must contend with the heating you put into the home.

Tips for Waste

Most household waste accumulates in the kitchen where food is prepared and where shopping is unpacked and unpackaged. There is an abundance of possibilities for 'doing something' to reduce waste.

Waste is at the heart of environmental problems: if conserving resources is 'heads', reducing waste is 'tails'. Waste is squandering the earth's resources and is causing an immense disposal problem because of the quantity, durability and often dangerous nature of the materials we use and throw away. In the home the amount of waste we produce is in proportion to the amount of goods we bring in.

Waste has been 'invented' by the throw-away society of the affluent world. In Nature there is no waste; everything that is discarded, such as falling leaves, fruits, branches and everything that dies, is decomposed and recycled into new life. In many developing countries, there is similarly hardly any waste for everything discarded by someone is transformed and re-used by someone else. The wealthy countries are only beginning to accept re-using and recycling as a basic way of doing things, both in industry and on a personal level. We have largely achieved our wealth by converting the world's resources into waste, a one way flow in which the lifespan of resources as useful products has become very brief. For the sake of the future of the planet we must begin to see things differently and to take Nature as a model.

WHAT YOU CAN DO

You can help by REFUSING

- *Refuse excessive and unnecessary packaging.*

- *Refuse plastic bags and remember to take your own!*

- *Refuse to buy poorly made goods that obviously won't last.*

- *Refuse to be convinced that something can't be mended.*

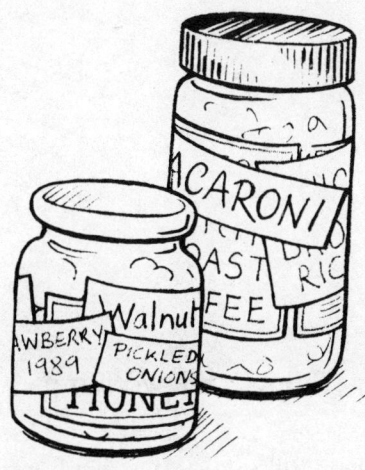

You can help by RE-USING

- *Buy drinks in RETURNABLE bottles where these are still available – or make your own drinks (see page 12).*

- *Save good packaging for re-use – paper, boxes, bags, bubble film.*

- *Save good polythene bags; all charity shops, jumble sales, bring and buy sales are glad of your surplus.*

- *Re-use aluminium foil, carry-out trays, etc., instead of buying new foil.*

- *Save scrap for schools, children's scrapstores, for making Christmas decorations and for children's rainy day activities.*

- *Big plastic bottles and white plastic squeezy bottles can be useful for the garden (see Young People page 42).*

- *The best way of re-using is to get something mended that would otherwise have been thrown away.*

Recycling and finding ingenious things to do with scrap will certainly help reduce waste and conserve resources. But the most effective action of all is not to produce waste in the first place. Reducing household waste is a convincing way for individuals to show that when they say they want to do something for the environment, they mean it.

You can help by RECYCLING

- *Taking all non-returnable glass bottles and jars to a BOTTLE BANK. Any broken glass can be put in a bank. For example, carefully place a broken pane into the middle of several sheets of newspaper, wrap it and gently tread it into small pieces. At the bottle bank, use the paper as a funnel: don't handle the broken glass with bare hands.*

- *Collect small aluminium pieces (foil, trays, bottle and carton tops, can rings) for Oxfam.*

- *Collect metal cans, flatten them and take to a can bank. Big aluminium articles can be sold to a scrap metal merchant.*

- *Take waste paper to a paper bank or save it for a charity.*

- *Promote manufacture of recycled paper by using it, especially items used in big quantities like toilet paper and kitchen roll (see Living Room page 38).*

WASHING UP and DOWN

'Spotless!' 'Sparkling!' 'Snow-White!' 'Germ-Free!'

Our zeal for cleanliness costs the environment dearly! These are some of the costs:

Costs in resources

Cleaning products and their plastic bottles are manufactured from mineral oil; plant oils and animal fats are used to make soaps. Chemical perfumes and colouring are added to both.

Costs in pollution

The valuable, mostly irreplaceable, resources used to make washing and cleaning products, bio-degradable or not, finish up in the sewage system, rivers and sea.

Costs in water

Dish washing, clothes washing, household cleaning, car cleaning, uses up gallon upon gallon of water, every drop of which has been purified to drinking water quality.

Costs in energy

Most washing water is heated. Spinning uses further energy to extract the water; tumble driers still more.

Costs in wear and tear

Clothes are often ruined by careless or excessive washing – shrunken woollies, run and faded colours, melted synthetic fabrics (by ironing too hot!). Scouring powders shorten the life of household surfaces.

WHAT YOU CAN DO

To Save Resources and Reduce Pollution

- *Use 'kind to the environment' detergents, cleaners and fabric softeners which are manufactured from plant oils, but use even these sparingly and only when necessary. Although more expensive as yet, they are very concentrated and don't cost more if used sparingly.*

- *Here is a bold step – don't use detergent at all if dishes aren't greasy such as after light breakfast, cups of tea and coffee, sandwich lunch, tea-time. Break the detergent habit!*
- *Here is an even bolder step! Use detergent (or fabric softener) in the washing machine only every second wash, even less often for once-used sheets and towels.*
- *Use bleaches and disinfectants with great restraint (see Bathroom page 26). Choose non-chlorine bleaches.*
- *Avoid using ALL AEROSOL cleaning products, even those that claim to be safe for the ozone layer. None are 100% safe. Spraycans are an extravagant way of packaging..*

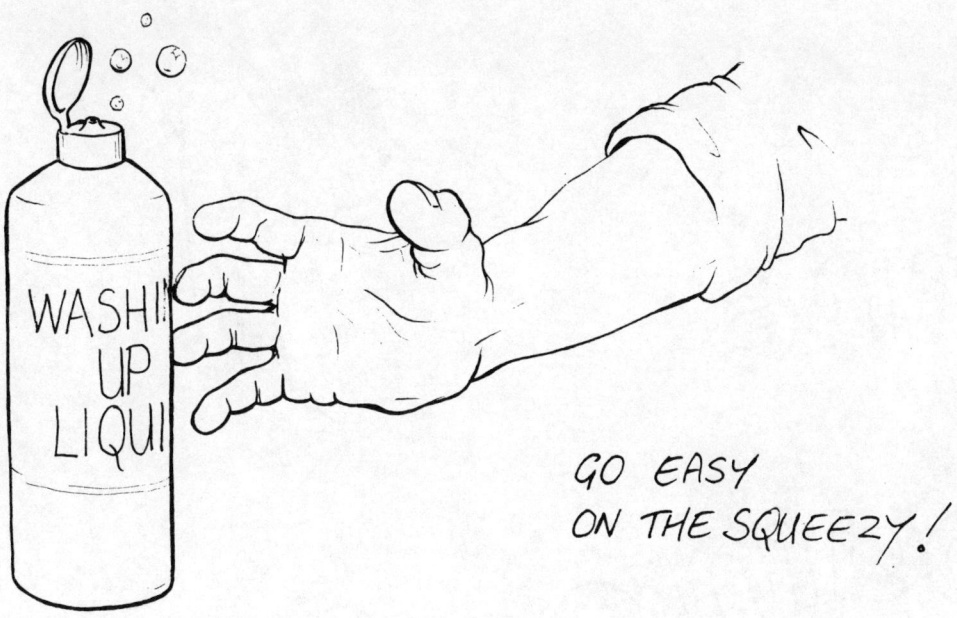

GO EASY
ON THE SQUEEZY!

To Conserve Water and Energy

- *Wait until you have a full load of either dishes or clothes before washing.*

- *Use hot water in a sensible order: lightly soiled articles first, heavily soiled articles last, finally the floor!*

- *Upgrade the insulation of the hot water tank. A D.I.Y. fitting jacket is best and pays back its cost very soon. Meanwhile, till you get round to fitting one, wrap old coats or blankets around the tank.*

- *Set the hot water thermostat to a reasonable temperature. The hotter the water the faster it cools, so it will cost you more to maintain the tank at a high temperature.*

- *Use only cold water for rinsing (clothes, dishes, cars). Rubber gloves will protect your hands from cold water (but caution, they will equally protect from excessively hot water and so you could end up wasting instead of conserving heat energy).*

- *Dry clothes out of doors if you have the possibility of doing so. Wind power will do the ironing for you!*

- *Save holed rubber gloves for gardening (see page 54).*

the bathroom

THE BATHROOM

Bathroom and toilet are close runners-up to the kitchen in the number of small changes you can make to help the environment. Many kitchen ideas apply also to the bathroom – how to use less water (and still stay clean!), how to save energy, how to avoid adding yet more to water pollution. The subject of animals and animal products crops up again – should one use cosmetics containing products from rare animals, indeed any animals; test toiletries and cosmetics on animals? Natural sponges; should one use them?

Conserving Water

We aren't at all water shy in the bathroom. We use water lavishly – every year, each person uses about 45,000 litres (an average room full). Every drop has been purified to drinking water quality. A lot has been heated at considerable cost to both

What? Save Water?!

pocket and the environment, is used briefly once and then emptied, carrying its valuable heat down the drains. If you reckon in the water used on our behalf by industry, agriculture and public

services, each person uses more than a swimming pool full each year! We are now using water faster than reservoirs can replenish themselves. If we don't restrain water consumption, we will have to build more reservoirs which will flood more valleys, destroy more wildlife habitats and disrupt the lives of more people. Without stinting ourselves there are a number of things we could do –

WHAT YOU CAN DO

- *Get leaking taps repaired now (don't put it off till water is metered!)*

- *Check that pipes in the loft or outside are protected against freezing. Don't risk the damage, expense and inconvenience of burst pipes. On top of damage done to your home, burst pipes waste a lot of water.*

- *If your bathroom has a shower (using 2 – 5 gallons) use it rather than the bath (using 15 – 20 gallons) for everyday. Luxuriate in a bath once in a while!*

- *If you wash hands under a running tap (we all do it!), don't turn the tap on full. A push-on spray fitment on the tap is nice to use and economises on water.*

- *Lastly, need one flush the toilet (using 2 – 3 gallons) after every little pee (strictly within the family, that is!)?*

Cutting Pollution

Every home can boast of two luxuries that we take totally for granted – plug holes in our sinks and the flushing toilet. That we don't need to worry about getting rid of our wastes or what happens to it outside our homes is a luxury indeed.

One of the most valuable of all the 'wastes' we produce is sewage. The sewage system, which was mainly built in the nineteenth century, was not planned to separate human sewage from toxic industrial effluent. A lot of sewage is therefore contaminated and cannot be used for its rightful purpose as agricultural fertiliser. So this valuable resource is degraded into waste to be got rid of, often at grave ecological cost to marine life, through dumping or discharging it at sea. Many bathing beaches around Britain are polluted with sewage and are a hazard to health.

Next time you are out shopping note the sheer quantity of bathroom toiletries and cosmetics sold in supermarkets or chemists shops. The whole lot is heading for the drains and will ultimately pollute rivers and seas. Pollution from bathroom toiletries is small compared with pollution from big sources like industry, agriculture or our own sewage, or even domestic washing machines, but that is no reason for doing nothing about it. As most toiletries are manufactured by the chemical industry, we are helping to reduce pollution at the industrial stage if we use less. Minimising the amount of bathroom toiletries we add to the brew in the drains would be helping to give back to sewage its former status of A1 resource.

WHAT YOU CAN DO

- *Could you cut down your use of soap, shampoo and other bathroom products? Could you wash with plain water as a rule, with only a touch of soap? Most people's skin and scalp would appreciate degreasing less often.*

- *Ban aerosols from the bathroom, even those with propellants that claim to be ozone friendly. None are completely harmless. All spraycans are a wasteful form of packaging and can't be recycled.*

- *Many products which have been until recently only available in spraycans are now packaged in plunger (trigger) operated spray bottles. These are very 'environment friendly' gadgets and can be re-used for many different purposes.*

- *Avoid putting a lot of disinfectant or strong bleaches into the toilet. A better way to keep everything fresh and hygienic is to give a few puffs into the corners of the toilet, basin, bath or shower from a trigger spray bottle filled with diluted bleach or disinfectant. This way you kill germs only on the spot, whereas pouring full strength bleach or disinfectant straight from the bottle kills germs indiscriminately in the sewage system where they do a valuable job. Label the spray bottle 'bleach' and keep out of reach of children.*

- *Whenever possible buy 'generic' medicines (marked B.P.) rather than 'proprietary' brands (with trade names). B.P. medicinals are the basic form of the same drugs. They are cheaper, more simply packaged and without added extras. Production, advertising and marketing of proprietary brands by the pharmaceutical companies is very wasteful and creates more rubbish and pollution.*

- *Just a penny worth of an idea – instead of using bath salts or other chemical perfume, give your bath water fragrance with fresh herbs or used lemon rind tied into a piece of muslin (cut from old tights).*

Energy Saving in the Bathroom

- *Economical use of hot water saves the most energy. Showers use much less than baths. Make sure to lag the hot water tank well (see page 23).*

- *Tape on insulating material round the outside of the bath.*

- *If you are aiming for top marks for simple energy conservation, what about this idea. Get a piece of bubble film (from packaging, or new from garden shops) of a size that generously fits the inside of the bath. Float it on your used hot bath water so that it covers the water surface properly. Once covered, the hot bath water will warm and dry out the bathroom, without causing any damp. Why heat the drains?*

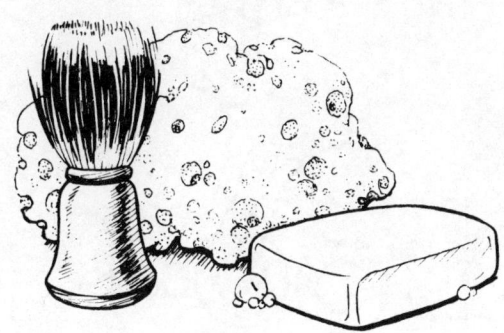

Beauty Without Cruelty

'Pour être belle, il faut souffrir' (to be beautiful you have to suffer!). But nowadays who is it that does the suffering? At the time when the phrase was coined it was people. Fashion and custom expected women, and men, to wear unhealthy, uncomfortable and unwashable clothes and hair styles. It seems incredible that only two hundred years ago women were setting their hair with lard and powdering their faces with poisonous white lead. Today people's health and comfort comes first and it is animals who are often made to suffer on our behalf. Cosmetics and toiletries must cause us no discomfort or ill-health, so test them on animals first! Many years of campaigning on behalf of animal rights has made the public aware that such testing is indefensible for products which are not necessities of life.

Many cosmetics and perfumes are made with animal products. Lanolin, which is the basis of many creams and ointments, is a fat obtained as a by-product of the wool industry and is in plentiful supply. Some expensive perfume ingredients however come from rare and threatened species, animals such as the sperm whale (for spermaceti), the basking shark (for squalene), the musk deer (for musk). There should be no market whatsoever for such cosmetics, so as to discourage hunting these animals.

WHAT YOU CAN DO

- *Before buying any cosmetic or toiletry, read the label and where possible choose one that states plainly 'THIS PRODUCT HAS NOT BEEN TESTED ON ANIMALS'. If this message is not on the label, assume that the product HAS been tested on animals. Many of the major cosmetic companies now market 'cruelty-free' products. If in doubt, ask before you buy.*

- *Avoid the more expensive and exotic cosmetics, especially perfumes, because these are more likely to contain rare animal ingredients.*

Animals and Plants in the Bathroom

Today's SHAVING BRUSHES are normally made out of pig (hog) bristles instead of badger bristles as formerly. But as a precaution, check the label and refuse badger bristle brushes. SPONGES are the homes of simple animals. They aren't rare but gathering them in quantity disturbs the ecology of the sea bed.

LOOFAHS are the fibrous insides of the dish-cloth gourd, a plant like a marrow – very environmentally sound scrubbers!

SPIDERS have a way of slithering into the bath – help them to climb out. What! Rescue spiders? Yes, please do; they are invaluable creatures who help to control bothersome insects.

Always use recycled loo paper

THE BEDROOM

Here we mainly consider fabrics and clothes – is it better for the environment to use materials of natural or synthetic fibres? What is the difference to the environment of buying new clothes or clothes that are 'new to you'?

A Few Facts About Fibres

COTTON is an important cash crop grown in most developing countries with hot climates and also in the southern states of the US and Russia. The cost of growing cotton is high, in terms of what it does to the soil and the people who grow it. To produce a good crop the plants need rich soil and plenty of fertiliser. They must be well irrigated and this causes many health and environmental problems. Cotton fields are sprayed very frequently because the plants are so prone to disease, thereby exposing people and wildlife to poisoning from powerful pesticides, many of which are banned in Britain. Half of all the pesticide used in the developing world is sprayed on to cotton.

In the US in the 1930's cotton growing helped to turn the fertile lands of the southern states into the infamous 'dust bowl' and cotton growing is still a major cause of soil erosion in the US and elsewhere in the developing world. Because the world price of cotton is falling, growers are forced into taking short cuts in cultivation which put themselves and the soil at even greater risk. Cotton does not have a good environmental record. However, some farmers in California are now producing good crops of organically grown cotton, showing that it is possible to grow cotton in a way that is safe and does not degrade the soil.

WOOL, LEATHER, FEATHERS and SKINS: the effects on the environment of producing animal fabrics are much the same as for producing meat (see page 11): sheep do extensive ecological damage to hill country whether they are raised for wool or for meat; animals are hunted to extinction whether for fur or food.

But using animal fabrics mixes ethical issues into the witches cauldron of environmental problems created by our need for clothes. Most wildlife organisations are campaigning for a ban on trade in leather, feathers or skins of rare and threatened animals. Others argue that having commercial value is the only way to protect rare species from overhunting and extinction. Some people (vegans) will object to wearing any fabric of animal origin. If one day in the future, we all reduce our consumption of meat and animal products and raise humanely a much smaller number of animals, we will also have less animal fabrics. We will have to manage with fewer pairs of shoes, make leather coats and handbags last longer, unravel and re-knit the wool from old sweaters, have fewer feather filled cushions, re-use old horn buttons.

So, might SYNTHETIC fabrics be a better

Cotton

choice? Synthetics are manufactured from oil and coal using a lot of energy and creating toxic waste. Toxic waste from the chemical industries of Britain and Europe ends up in the North Sea, either by direct dumping or carried there by polluted rivers so that it has become one of the worst polluted seas in the world. Synthetic fabrics are light, strong and durable; these very advantages sometimes create serious problems. *Do you dream of the sea? Do you have net curtains at your bedroom window? They are (probably) made of synthetic fibre. They do no harm to anything, but they are the small relations of mon-ster nets, called drift-nets, used by many nations for industrial fishing. Drift-nets form barriers across the oceans up to 40 miles long. They are so strong and durable that years after they have been abandoned, they still entangle and strangle big fish like tuna, or mammals like dolphins and small whales, creatures who could have struggled free of rotting, natural fibre nets – a nightmare by courtesy of synthetic fibres* (see also page 13).

It is difficult to make a choice between natural and synthetic fabrics on environmental grounds since both have advantages and disadvantages. As with all consumer goods,

the best way is to be sparing with new purchases whatever the material. Weighing things up, the scales come down in favour of making natural fabrics the first choice (with some exceptions) because:

1) They are renewable resources which COULD be produced in a sustainable, environmentally-friendly and people-friendly way.

2) They are more satisfying to handle and work with and are more comfortable (and generally healthier) for most people to wear.

3) Discarded natural fabrics can be better processed into useful materials, such as industrial wipers and carpet underfelt, than synthetics.

4) Natural fibres rot away when finally discarded and so don't add to the rubbish problem.

WHAT YOU CAN DO

- *Buy the best quality that you can afford, whatever the fabric. They look better for longer. Cheap clothes exploit people, resources and the environment the most.*
- *Don't take risks with washing, cleaning and ironing. Follow the instructions! Too frequent washing shortens the life of natural fabrics.*
- *Get shoes mended in time. Give old shoes and handbags a new lease of life with good quality shoe dye. It works!*
- *Choose synthetic fabrics and fillings where these are obviously better suited, for example for bedding which needs washing and for carpets which need durability.*
- *Save good parts of garments for linings, mending, making patchwork and for rag weaving. Exchange remnants with friends. Many charity shops are glad to have buttons and good zips.*
- *Make patchwork cloth for bedspreads, cushion covers, curtain, duvet covers, cloth bags, etc. You don't need to labour over tiny pieces; a patchwork of big pieces can be most effective.*

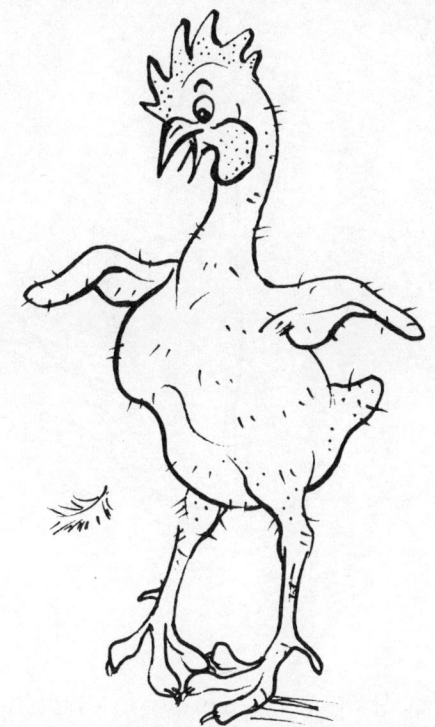

- *Explore your High Street charity shops for 'new to you' garments before deciding to buy quite new (see next page). 'Recycle' your used clothes and furnishings by selling or giving them to a charity shop. Oxfam sends unsaleable garments for processing into felt and other products.*
- *Or take rags and worn out clothes to a textile bank, to be found at local civic amenity dumps and some scattered collecting points.*

Keeping warm with Ducks and Duvets

Before duvets became popular we had eiderdowns. The most luxurious of these were filled with down from eider ducks, hence the name. The down, which the eider ducks pluck from their breasts to line their nests, is gathered after the nesting season is over – a small but sustainable harvest. However today's pillows and duvets are usually filled with feathers or down which is a by-product of the poultry industry. Using its by-products helps to keep this inhumane and polluting industry going, so until poultry farming is reformed, duvets and pillows with synthetic fillings are better environmentally. They also have the advantage of being washable.

Switch on the **electric blanket**? Why not! It's more economic to heat the bed than the bedroom! (Always follow the instructions for the use of electric blankets).

New to You

fashion, we can be free to experiment with different styles. Charity shops and nearly new shops put this within everyone's means. Buying or giving clothes this way helps to conserve resources and to slow down the degradation of the environment. Stop to look in next time, before you decide to buy new.

The idea of buying nearly new clothes has become quite accepted. More and more people are discovering the advantage, and the fun, of rummaging through the interesting variety of garments on sale in charity shops. Charity shops have become so popular that there is today hardly a shopping street in Britain without at least one. Charity shops offer something for everyone, which is not really surprising since it is 'everyone' who provides what's for sale. They are friendly and welcoming shops where no one will hurry you or pressure you to buy.

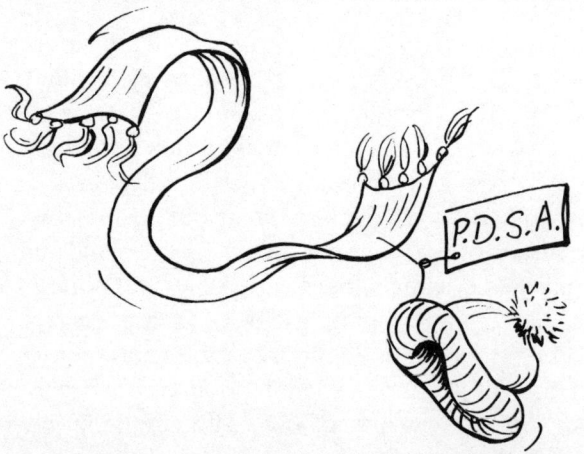

Once upon a time we dreamed of manufacturing clothes so cheaply that nothing need last for more than one season. We even tried paper clothes, the ultimate in a throw-away society. Today, because we are much more relaxed about

Designer Clothes and Accessories by Traidcraft

If you want something new and special, your purchase can still help people and the environment. Traidcraft (see page 80) offers a unique collection of clothing for both sexes (not children) made in the developing world with traditional skills but in styles that fit in with our way of dressing.

To quote from their Clothing Collection catalogue:
'Traidcraft clothing is not mass produced. Garments are made individually, often using handloom cloth. The maximum quantity made of any garment is 600, so each garment is quite exclusive. Traidcraft clothes are not just beautiful, they are part of the business of people centred trade.'

Write to Traidcraft for a catalogue and the name and address of your local Traidcraft representative.

hall,

stairs and landing

HALL, STAIRS, LANDING

This may seem an unlikely part of the home in which to find ways to help the environment, but even here there are quite a few worthwhile things to do. Draughts may whistle through the hall wasting heat, junk mail may pile up behind the front door wasting paper, a light may need to be left on for long periods, doormats may need to be replaced; a use for snow is proposed and some environmentally helpful ways to furnish the hall are suggested.

Keeping the Heat In

Draughts cause the biggest heat loss in the home; the air in a draughty entrance hall changes many times over in an hour on a windy day! Because the entrance hall, staircase and passages connect all parts of the home it is particularly worth draught sealing here. You will save more heat by draught sealing than by expensive home improvements like double glazing. However, it is possible to seal so effectively that you stop all ventilation and get problems with damp, especially if you heat with paraffin or bottled gas.

WHAT YOU CAN DO

- *Draught sealing is best done on a windy day. On a still day it is impossible to know whether you have done the job effectively.*
- *'Double glaze' with a sheet of polythene any windows that don't need to be opened.*
- *Hang floor length curtains (old bedspreads do well) over draughty doors, if you can't seal them, or to cut off a cold corridor or passage. Light weight curtains will hang better and seal off draughts better if you put a weight in the hem: thread in a length of chain or a row of pebbles.*
- *If your house is damp, there is a simple, inexpensive ventilation system that works together with draught sealing, write to:*

David Hugh Stephens, M.Sc.
Victoria House
Bridge Street
Rhayader
Powys LD6 5AG

David Stephens is a conservation engineer, developing simple, environmental technologies. He believes that keeping a low level of heating going continuously during the cold months, instead of heating up a very cold house each evening, actually saves energy and prevents damage from damp.

Junk Mail

Cold air isn't the only uninvited thing to blow through the letterbox! What about that stream of advertising mail (known as direct mail, junk mail) that pours into most households? How much of this is of interest to you? How much goes straight into the wastepaper basket?

WHAT YOU CAN DO

- *Write to ask to have your name removed from a firm's mailing list. It usually works.*
- *Write to:*

MPS will send you a form on which you enter the names of persons in your household who do not wish to receive direct mail (advertising, etc.). This information will be passed on to those businesses who subscribe to MPS. They will remove your name and address from their lists. This should stop quite a lot of unsolicited advertising from reaching you, but it will also stop appeals from charities who subscribe to MPS services.

In addition you can write to the Advertising Standards Authority for a leaflet called 'Personally Yours' which explains ways in which you can have some control over direct mail.

Furnishings

The hall, stairs, landing and passage are good places to use energy saving light bulbs, especially if a light is on for long periods. If every home replaced just one 100w bulb with a low energy bulb, one power station could be shut down. And, if this light is on for eight hours per day, it should last 3-4 years – quite an advantage where a stair light is hard to get at.

Before shopping elsewhere for furnishing items for the hall and stairs, look into an Oxfam shop or get a Traidcraft catalogue (for address see page 77). Both have many useful and attractive items such as rugs, doormats, wall hangings, mirrors, hanging baskets, hand made in small workshops in developing countries.

Snow-Clean

Here's a traditional, effective and invigorating way of cleaning rugs:

> 'When the snow lies round about,
> deep and crisp and even.
> Throw your rugs and carpets out,
> frost and snow will clean 'em.
> Heap the snow on, brush with might,
> See the colours changing,
> Snow turns grey and rugs turn bright
> And chem-i-cals you're sa-a-ving'

(Sing to the tune of Good King Wenceslas, any time there's a good covering of snow!)

THE LIVING ROOM

The living room is a big user of trees; timber for furniture and fittings, wood pulp for books, newspapers, magazines, personal computer paper, writing materials, wall paper, lampshades, pictures, frames, etc.

If all the timber used in the world in just one year was piled on to Birmingham, it would bury the city under a mound the height of a 10 storey building. 27 million tons were imported into Britain in 1987 as timber and wood products; the quantity goes up each year. Picture two heaped-up (big!) car trailer loads of timber; that is what each of 56 million people in Britain uses up every year and the loads are getting bigger. Much of the timber we use is cut in the shrinking tropical forests; most of the pulp we import comes from temperate, coniferous forests, few of which are managed sustainably. Our insatiable demand for wood and wood products is destroying the greatest and most extraordinary forests in the history of the planet.

Although some timber is grown in Britain, mainly softwoods for pulp, most of what we use is imported. Over one million tropical hardwood doors alone are imported into Britain each year. And this is just what is visible. Out of sight is the timber used for the frameworks of upholstered furniture and fittings, and construction timbers for house building. Outside our homes timber is used in industrial and public buildings, for transport structures, for boat building. We are the largest importers of tropical sawn timber in Europe. Our demand is so great because so few hardwood trees are grown for timber in Britain. Look round your living room: tables, chairs, television cabinet, sideboard, lamp base, picture frames, fruit bowl. You will be lucky to find even one item made from a British hardwood! Why is this so? Because most of Britain has long ago been turned to agriculture to feed its oversized population and it is easier to use up other countries' forests than to plant and manage our own sustainably.

Then there is a stream of paper flowing in and out of most living rooms: newspapers, magazines, writing paper and envelopes, Christmas and birthday cards, notebooks and diaries. Perhaps you have a personal computer that devours paper by the yard? Every year, each person 'uses' paper equivalent to two whole trees: a lot of this is used on our behalf by government and industry. Forests of conifers are felled for each week's New York Sunday Times; British Sunday papers are not far behind! Even a proportion of tropical wood is now converted to pulp, mainly for cardboard.

It will one day seem incredible that in the late 20th century, such wonderful living beings as trees were still being pulped into products as ephemeral as paper – just as outrageous as turning whales into pet food and machine oil. Far fewer trees worldwide will need to be cut down, far less of our countryside will be blanketed in spruce, when we develop extensive recycling of paper. Many businesses and government departments are changing to recycled paper, but it is the everyday use of recycled paper by people that will make the big difference.

WHAT YOU CAN DO

The destruction of the tropical forests is so dire, so threatening to the balance of life of the entire planet that we must immediately stop the excessive use of this timber. When buying wood or articles made of wood, the order of choice should be – **first**, old salvaged wood: **second** British (or European) hardwood: **third**, softwood (pine): fourth, tropical hardwood from a plantation: **never**, tropical hardwood from virgin forest (but at present there is unfortunately no sure way to tell virgin from plantation tropical timber).

GIVE PRESENTS made from British hardwoods (bowls, boxes, lamp bases, bread boards, cheese boards, chess sets, toys, picture frames). This encourages local forestry and timber businesses, crafts people and creates jobs. Only buy items of tropical woods from a Fairtrading shop or catalogue (OXFAM, TRAIDCRAFT, GREENPEACE, ACTION AID, see page 80).

WOOD/TIMBER

POSTPONE buying new furniture, having a new fitted kitchen, building an extension to the house, while you look into the possibility of obtaining timber produced with less environmentally damaging impact (ask advice from Friends of the Earth, page 78). Old secondhand timber is often of a higher quality than anything new – a measure of how forests worldwide have deteriorated.

REPAIR furniture instead of buying new, especially old items of quality. You are conserving forests if you spend on repairs what you would have spent on buying new.

MAKE AVAILABLE to others any unwanted furniture; don't throw out or burn (unless badly woodworm infested). Some charities take furniture and appliances and some local authorities run workshops and make available repaired articles to people in need.

PAPER

BUY only recycled paper. By doing so you promote paper recycling. Choose off-white paper for most purposes. Watch out for the EC label (later in 1993) which will tell you that the paper contains maximum recycled (post-consumer waste) fibre.

TAKE waste paper to a paper bank and offer to take your neighbour's as well.

WATCH the news on telly (or listen on radio!) and buy fewer papers each week; share newspapers and magazines with friends.

SAVE Christmas cards. Cut off the message and re-use the picture as a notelet.

SLIT envelopes with a knife so that they can be re-used with labels from charity shops or from a plain perforated roll (stationers).

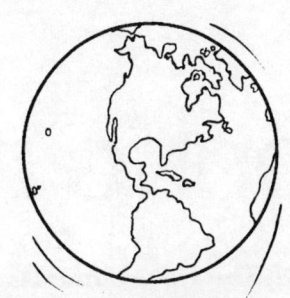

young people

YOUNG PEOPLE

What Have Pets To Do With The Environment?

Keeping a pet is something most children dream about and many lucky children are given one at some stage. It goes without saying that every animal deserves to be looked after in the best possible way with the right sort of food (though not too much!) and plenty of exercise. But there is another side to keeping a pet; in doing so you have taken on twin responsibilities, for the pet itself and for what the pet may do, directly or indirectly, to the environment.

Choosing a Pet

Animals that have been bred domestically or in captivity for many generations have become adapted to live that way. These animals make the best pets: dogs, cats, guinea pigs, hamsters, gerbils, mice, budgerigars, rather than more exotic and fussier species. Choose hardy breeds rather than special highly bred ones which are often delicate and may have special requirements.

Never buy an animal that has been taken direct from the wild because this encourages poaching and illegal trading in wildlife. Most exotic birds and animals don't survive transport from their country of origin, so for every tropical bird, turtle or monkey that arrives still alive, dozens will have died on the way.

Always buy from a reputable pet shop and ask lots of questions about breed, hardiness and where an animal has come from, before you fall in love with it.

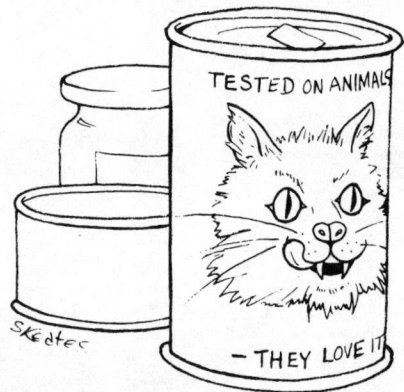

Pet's Food

When choosing a pet you should think ahead to what food it will need – meat or vegetable? If you yourself are a vegetarian (or near vegetarian), and many young people are, does it make sense to keep a meat-eating pet?

Dog and cat food is a multi-million pound industry. More tins of Whiskas are sold than any other single grocery line! Dog and cat food is partly made out of waste from the meat industry but many brands also contain good quality meat and fish imported from developing countries that should be feeding people in those countries. Our dogs and cats can pay more for their food than many people in the developing world can pay for theirs. Worse still, some pet food contains meat from protected animals such as whales or kangaroos (this way European dogs become predators of the kangaroo!). A meat-eating pet is inevitably part of the world meat business which is one of the most environmentally damaging activities of agriculture (see FOOD page 11).

Pet's Freedom

More directly observable is the effect of pets on wildlife – dogs and cats in particular. Dogs love being given freedom to chase around in parks or the countryside. But keep an eye on your dog, as you enjoy watching him enjoy himself. Wildlife can be seriously disturbed by an exploring dog, even if it is trained not to chase or hunt. Roaming cats are also disturbing to wildlife, specially to birds in the nesting season. Listen to the alarm calls of birds when a cat is on the prowl!

There are many ways in which people's oldest and best friends can come into conflict with wildlife and the environment. It isn't a reason for not keeping dogs and cats but it is a reason for there not being any unwanted ones around. Stray cats and dogs do a lot of harm to wildlife, especially if they breed in the wild (become feral). Making sure that a pet doesn't stray, while allowing it freedom to exercise, is a responsibility every pet owner has towards the environment.

Here are some other ideas of how you can help the environment:

HELPING BY COLLECTING THINGS

Many things of no use to one person can be of great use to some one else. Your parents may groan at the thought of encouraging the magpie instinct in you! However, this sort of collecting is different because you will be passing things on as you collect; some very useful things take up very little space but tidiness must be the order of the day!

Here are some things worth collecting:

Stamps – British and foreign: *most charities collect these to raise funds.*

Postcards and Christmas cards: *some card collections can raise money, others can be useful for schools, children's wards. Christmas cards can be re-used as notelets – stick them shut and write on the back.*

Returnable bottles: *you could run a Bottle Return Service for people who aren't able to take bottles back themselves or who don't know which shop takes back which bottles. Look out for undamaged returnable bottles lying around (parks in town, roadside hedges in the country) and raise some funds for a good cause at the same time as reducing litter.*

Non-returnable bottles: make the round of your neighbours to collect empty bottles and jars to take to a bottle bank.

Paper: if your Dad is taking waste paper to a 'bank', run round to the neighbours and offer to take theirs.

Aluminium: if you can collect enough big stuff, like old saucepans, kettles and other aluminium scrap, you could make a little pocket money (and donate some to a good cause) by selling to a scrap metal merchant. Oxfam collects the smaller scrap (foil, carry-out trays, milk bottle tops, can rings).

Tins: collect empty tins, especially pet food tins; wash and flatten them. Under 10's can do a 'workout' of excess energy by jumping on them.

Plastic: plastic recycling is on the way in, so if you can find a collecting point, you can start collecting plastic waste and considerably reduce the volume of rubbish in your dustbin. Charity shops and jumble sales are always in need of clean plastic bags.

HELPING BY MAKING THINGS

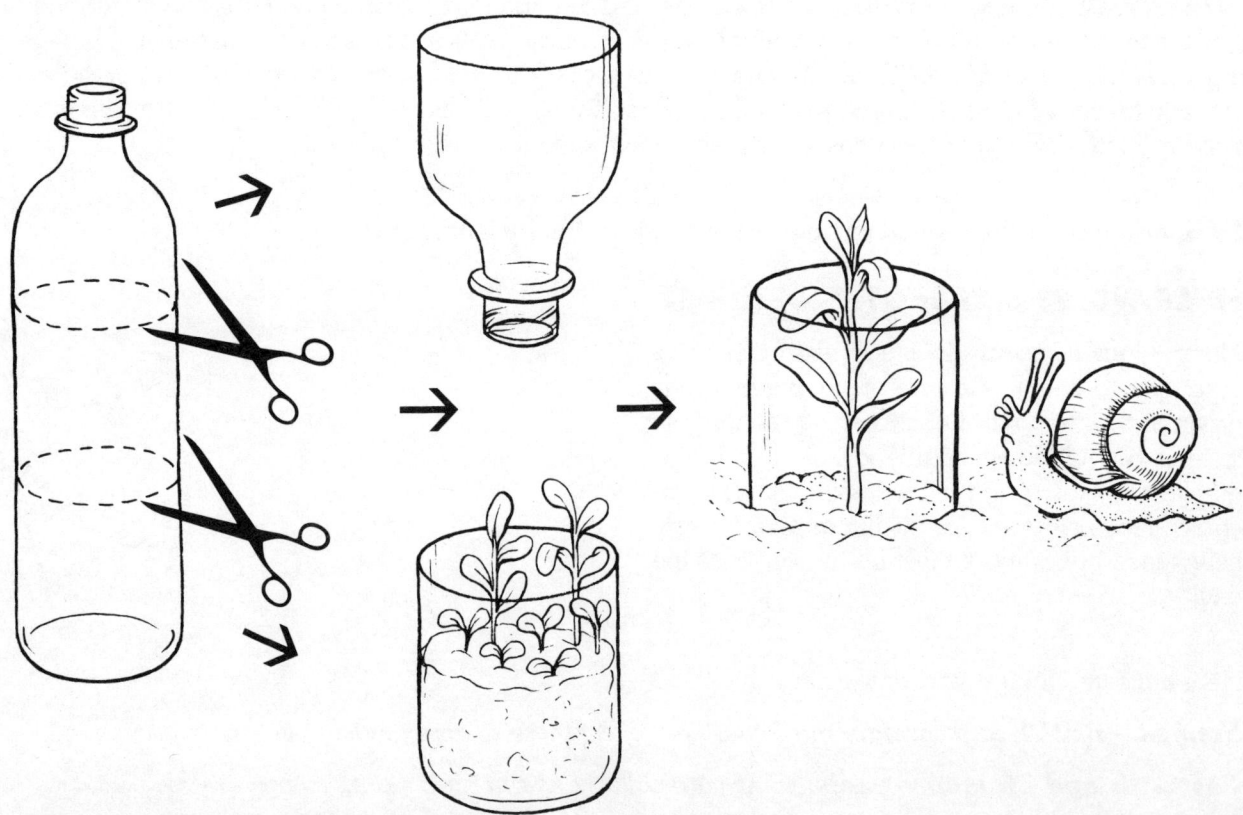

HELPING BY MAKING THINGS

From plastic bottles: Cut into three pieces. The bottom piece can be a plant pot, the middle piece can protect seedlings from birds and slugs (see picture), the top piece makes a serviceable funnel for the kitchen or garage.

42

From white squeezy bottles: Turn these into seed labels. Cut into 10 cm strips (names to be written on the plain white side with a waterproof pen). This makes a welcome present for a gardener as seed labels are expensive to buy.

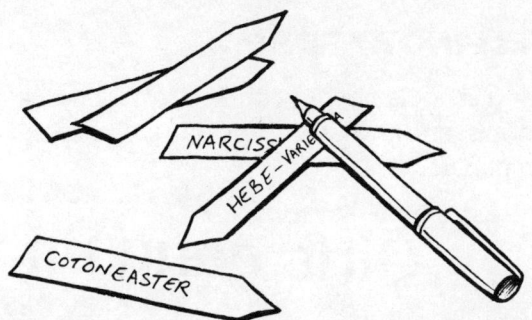

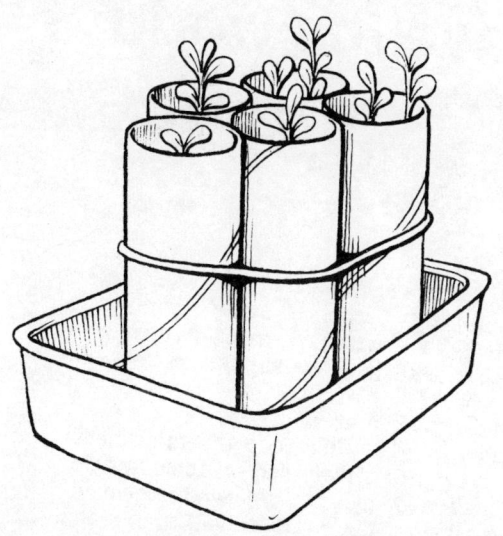

From toilet roll centres: Bundle a few together with a stretchy rubber band and stand them in a plastic supermarket food tray like those used for meat (a deep one is best). Fill the tubes 3/4 full with soil and plant one to four seeds in each. Water sparingly from the top; plant out the whole tube with seedling(s).

Safe scrap of all sorts is always welcomed by schools, nursery schools and children's SCRAPSTORES (is there one in your area?).

SAVE SAFE SCRAP · SAVE SAFE SCRAP · SAVE SAFE SCRAP
Try saying this fast, over and over!

although it is good to find lots of ingenious things to do with scrap, it would be even better if we **DIDN'T PRODUCE SCRAP** (waste) in the first place!

HELPING BY GROWING THINGS

If you have a garden, even a small one, set yourself up to grow seedlings to supply your own or friends' gardens. Use (scrap) supermarket plastic containers with holes pricked in the bottom as seed trays. Try lettuce seedlings, cabbage, brussels sprouts, broccoli, sweet peas (start early) and different sorts of bedding plants. Growing good strong seedlings is not easy, so don't get disheartened if it doesn't work too well the first time. Place aluminium foil (used will do) as reflectors behind seed trays on a windowsill for an early start. Borrow a book on gardening from your library.

HELPING AT SCHOOL

Many schools now encourage lots of activities to help the environment. The pupils at these two schools, one in Scotland and one in America, are running environmental project that have made it into the papers!

Girls gang up to save the world
by Peter MacDonald

Emma

Karen

Katherine

What is unusual about this quartet of 11-year-olds is that they want to be for the world.

Catriona

What seems to have started as something vaguely interesting has become absorbing. They have a clubroom in an attic at Katherine's home where they have built up a library for study and each has developed her own speciality.

"We wanted to start a club," says Emma, "but we wanted to do something useful so . we started an ecology club because we thought it would be interesting."

Emma has her long-standing interest in animals, Katherine is concerned with general environmental problems, Catriona with marine wildlife and Karen with the rainforests.

These young reformers have impressed many adults, perhaps because they have done two wise things. First, they have taken the trouble to inform themselves about their subjects and, second, they have begun where they are.

Their latest project has been to interview the Scottish executive of WWF, who said afterwards: "I was so impressed with four young people who knew so much more than a lot of adults. They really grilled me."

Their school bears the marks of their effort - in a colourful display of posters and other material on conservation, in plans for a presentation at a school assembly and a sponsored walk in aid of the Worldwide Fund for Nature.

Oyster River Children Set Example
by Charlotte Kimball

'Trash Into Cash' is Oyster River School's motto

DURHAM - Oyster River school children are setting a fine example for their elders. They have been actively engaged in a recycling project that saves energy, reduces pollution, and helps curb strip-mining. Their drives to collect aluminum cans have resulted in less litter, and at the ame time raised money for a school project.

Over 118,000 Cans

Over 118,000 aluminum cans have been collected. Carroll and her committee met with students and explained the importance of recycling. They encouraged them to save cans at home. Students picked up cans saved by relatives, friends and neighbors they contacted. Businesses, restaurants and UNH dorms and fraternities and sororities provided other sources.

Newspaper publicity helped spread the word. In response to the advertising, a man in Rochester donated a garage full. It took three truckloads to cart them away. A real recycler, he felt it sinful to waste all that aluminum.

In addition to cleaning up the landscape, aluminum is a very energy intensive product to manufacture. The recycling of an aluminum can saves about 95% of the energy it takes to make a brand new can. Throwing away one aluminum can is the same as throwing away half a can of gasoline. Some UNH students did a recycling project and discovered the energy saved by recycling one aluminum can could power a radio for 51 days, and six aluminum cans can run a TV for 200 days. The aluminum that goes into a can is often worth more than its contents. Bauxite from which aluminum is obtained, is acquired by strip-mining, which means digging up more of the earth's surface.

Another example of the possibilities of such a recycling project is what happened in the small town of Enfield, New Hampshire. Students there collected 67,526 lbs. of aluminum in a three year project. Enfield Elementary School's 'Cans for Computer Corp,' grossed $17.540 and 75 students earned their own computers. Next year proceeds from their project will go toward refurbishing the playground.

why
on earth
?

WHY?

A Conservation Conversation

For an environmentally supportive 'green' lifestyle to take root and not fade away as a passing fashion, we must be clear about WHY changes are necessary. It is sometimes hard to make changes even in our own immediate interest, for example for our health; it will be harder still to make, and sustain, changes for the long-term welfare of the planet. Living as we do, in a well-off country, we are seldom aware of how we affect the wider environment, or are affected by it. We scarcely notice if the harvest is poor in Britain, or in one of its supplier countries, because food can be imported from elsewhere: if frost wipes out the coffee harvest in Brazil we grumble about a (small) rise in price but go on drinking coffee: it's not the plants in our gardens that die from our car exhaust gases, but trees in distant places. The invisible lines linking us with the environment stretch far beyond our horizons and are intertwined in an intricate cat's cradle of cause and effect. Because we get so few direct signals, we need understanding, backed by feeling, of our global environment, to sustain our motivation for change.

Naturally there will be doubts in everyone's minds: 'Can my small efforts really make any difference? Why should I inconvenience myself? Isn't looking after the environment the government's responsibility? Surely industry is the most to blame for pollution?' It is essential to discuss these nagging questions, which threaten to undermine our motivation, because the changes to lifestyle we will have to make, voluntarily now or out of necessity in the future, will have to be changes for always. We have come to the point where Nature's restraints will have to be replaced by our own, where all our activities must be for the planet, not against it.

Here are some likely questions and some possible, but by no means the only, answers.

Q. What difference can doing any of these little actions make? Why should I bother?

– Small changes will add up to sizeable effects because there are so many millions of us – a fact well known to the commercial world!
– Small changes prepare the way for bigger changes tha we will all have to make, sooner rather than later.

Q. Isn't it government's job to look after the environment?

– Legislation to conserve the environment can only work effectively if supported by people's goodwill. Getting involved by doing small things and understanding why creates this co-operation and can even lead the way for further government action.

Q. My friends think I'm silly to bother collecting up aluminium bottle tops, foil, cans, cluttering up my kitchen with bottles for a bottle bank, stacking up waste paper for collections, etc.

– Your friends are right; it IS silly! Local authorities should support recycling by providing separate containers for paper, plastic, glass, aluminium, other metals, and compostable wastes. Householders should get paid for sorted rubbish. Then it won't look silly any more! It's more likely to happen if we START by keeping bottle banks, can and paper banks filled and show that we, the public, are a step ahead in caring for the environment.

Q. Won't leading an environmentally supportive life-style turn us into misers and hoarders? Using less, re-using and recycling every scrap doesn't attract me! Why can't we go on living as we are used to?

– When you realise that the privileged 20% of the world (that includes us), use 80% of the world's resources and the gap is getting bigger, it is plain that we can't continue as usual for much longer. At present we have to make a conscious effort to re-use and recycle, but in an environmentally supportive society, doing this will be a matter of course. It won't then be thought of as a miserly lifestyle.

Q. Isn't conservation going to make for a dreadfully dull world? The exciting things in life all seem to be those that use up lots of fuel energy and resources and pollute the environment – spending 'lotsa' money, fast cars, space travel, exotic holidays, high-tech sports and entertainments, bright lights.

– We are going to have to wean ourselves from depending on these as sources of satisfaction and fulfilment and rediscover the human energy within ourselves. Human resources can easily provide all the thrills and spills (we should say 'thrills and skills') that we will ever need.

Q. Aren't we all going to look dreadful, wearing patched up hand-me-downs?

– If new clothes cost their real price, i.e. if the cost of degrading and polluting the environment and of exploiting people's health and labour were adequately paid for in the price of clothes, they'd be a lot more expensive. We'd look after them better, wear them longer and wouldn't find wearing nearly-new clothes at all unattractive.

Q. Isn't it being unrealistic to think that if we use substantially less energy and resources we will still be able to live comfortably?

– It depends on how much of the resources and energy we use so profligately we judge to be necessary for an adequate level of comfort and convenience. Improved efficiency could more than halve energy consumption worldwide without a drop in living standards. However, we will certainly have to go further than this. We could meet our resource needs sustainably (including the need to sometimes indulge ourselves!) by doing things in ways that make lighter demands of the environment – re-using, recycling, eating local produce, working with, instead of against nature. Makes sense, doesn't it?

Q. By adopting an environmentally friendly lifestyle, aren't we putting the clock back, going back to the days of the ox and cart?

– Making lighter demands on the planet does not mean turning away from technology; quite the contrary. It means using novel technologies that will enable us to conserve rather than exploit the environment. However living ecologically will mean developing the skills to do more for ourselves and each other, to lessen our near complete dependence on having everything done for us by industry.

Q. It's all very well for the better-off to be concerned about the environment but you can't expect the same from the less well-off: they buy little enough as it is and can't afford costly environmentally-friendly products.

– There are many actions to help the environment that have nothing, or very little, to do with spending money; many actually save you money. Because 'Green Consumerism' has attracted so much attention, it is easy to overlook those actions to help the environment that don't need money. There are lots of such suggestions in this book!

Q. I am concerned about pollution but now that I use environment-friendly (green) washing products, I don't have to worry about that one any more, do I?

– There is a danger that by emphasising environment-friendly features, manufacturers will give the impression that their products are 100% harmless. We could be tempted to use as much or more than before. However, ALL products, even the most environmentally friendly, have some impact on the environment and many green products are only a small step away from their non-green equivalents. Green products also use up energy and resources to produce, package and distribute. Safest for the environment is to use less, even of environment-friendly products.

Q. In the past, scientists have always managed to solve problems with new inventions. Isn't it likely therefore that new technologies are just around the corner which will rescue us from our environmental problems?

– This might have been true in the past but is most unlikely now. Now, with 5,000 million people on Earth, the ability of the environment to cope with our activities has become severely stretched. Even if it were possible to do so, inventing more mega-technologies to cope with the legacy of present mega-technology, will bring the environment to breaking point. The new technologies will need to be different; they will need to co-operate with nature and to take into account the connection between things. Their development and use will need to go hand in hand with changes in attitude – of the sort described in this book.

Q. Environmentalists say that we will help if we use less resources, buy fewer goods, reuse, mend . . . Will this not deepen the recession and throw more people out of work, both in Britain and elsewhere in the world?

– Most people are made redundant by automation and other cost cutting measures, necessary to keep industry and agribusiness going in the face of severe competition from other countries. Moving industries abroad to countries where labour is cheaper (exploited) and environmental controls are negligible is another major cause of unemployment here.

Many more people could be employed if we shifted towards more small-scale production. This would happen if we went for quality rather than quantity in our purchases, if we bought more handmade items, if we promoted durability and more production of spare parts for repairing things, if we replaced present chemical agribusiness with organic farming. There are many ways out of the present impasse if we are willing to make some radical changes to our attitudes.

Q. How do we know for certain that global warming is actually happening, that the extinction of a number of species matters, that North Sea seal deaths are due to pollution, that organic food is better than conventionally produced food? Don't we need to do more research to answer these questions before we take costly action?

– We always argue that the evidence is incomplete and that WE NEED MORE RESEARCH. Yet it needs only common sense, not research, to know that it is asking for trouble to put millions of tons of carbon dioxide and pollutants into the atmosphere each year, whether the consequences are fully understood or not. We know that all wild animals eat organic food and have a level of health and vitality that we cannot approach; do we need more evidence than this? We must assume that all pollution is guilty until it is (rarely) proved to be innocent. How much environmental damage could have been avoided had we acted years ago when the commonsense, but scientifically unproven, viewpoint was first voiced? Must we wait for disaster before taking action?

Q. It seems to me that some manufacturers are using the public's concern about the environment to promote their sales; messages about environment-friendly features are becoming a selling point. How can a shopper tell which products are really less damaging to the environment from those that are little different from the standard product?

– In principle, incorporating even one small environment-friendly feature is a step in the right direction. Shoppers are right to indicate to manufacturers their preference for such products in the hope that this will push them further in this direction. However, this may encourage producers and manufacturers to get away with doing just enough to appear environmentally concerned while avoiding making effective changes to the production processes. To deal with this problem, many countries have an eco-labelling system; the best known is the German 'Blue Angel' symbol. In Britain we have the Soil Association Symbol for foods grown 100% organically and the 'Conservation Grade' label for food which is produced in a kinder-to-the-environment way. Starting in 1993, there will be a European Commission Eco-label, awarded to the least damaging products. These labels are awarded to products which have been assessed for their environmental impact 'from cradle to grave'. Labelling will help the shopper, but there will surely also be a crop of phoney ones, so one will still have to apply discrimination!

the garden

HOME
SWEET
HOME

Skeates

THE GARDEN

Is your garden fit for wildlife?

There are close on 16 million private gardens in Britain. Garden land covers a sizeable area of the country. If all gardens were put next to each other, they would cover an area as big as a county like Norfolk or Northumberland. People lucky enough to own a garden, or have the use of one, have an anchorage to earth and a lifeline to Nature. You can strengthen the line – or sever it! – depending on what you do with your little piece of the planet.

Agriculture, commercial forestry and industrial development has greatly reduced the number and variety of wild plants and animals. Gardens can't replace wild nature but gardens can offer a far greater variety of food and shelter for wildlife than a barley or potato field. Above all, gardens in both city and country can be a refuge for wildlife and wild plants from the agrochemical warfare waged over so much of the countryside.

Why Have Wildlife in Cities?

A city that is fit for wildlife is fit for humans too. A built-up environment devoid of living creatures, plant or animal, will seem to most people to be a soulless and depressing place. Wild animals offer a special sort of companionship, keep us in touch with nature and save us from becoming too human-centred. Greenery refreshes the air, and the spirit, by filtering out noise, traffic fumes and dust. Trees actually improve the climate of a big city, if there are enough of them. Gardens make a network of green lanes along which wildlife from the surrounding countryside can move into the city.

You may be wondering why we should encourage typical city wildlife – do we want more pigeons? starlings? mice? rats? squirrels? foraging foxes? more slugs and snails? or, on the plant side, more dandelions, docks and stinging nettles? These are robust, adaptable species that thrive in the built-up environment. But cities also harbour plants and animals that are less common, less

bold. These grow in undisturbed open spaces such as railway embankments and other over-grown corners. Gardens will help to increase the variety and numbers of these species and help to maintain an interesting mix of plants and animals in cities. Grow plants that will make your garden attractive to a range of wildlife. Don't let the pigeons and sparrows have it all their own way!

To attract wildlife, a garden needs to provide **food, shelter** and **security**.

Food – every plant is food to some creature, though some offer far more than others. Wild or old-fashioned flower varieties often produce

skeates

more nectar, for bees and insects, than many modern plants bred for showy blossoms. Native trees and shrubs are better adapted to provide food and shelter for our native animals than ornamental and exotic species; oak trees support as many as 300 different species of insects. Although slugs and snails, and insect pests, are unwelcome in gardens, they are nevertheless an important source of food for birds. A feeding table can help birds survive a severe winter; they will repay you by gobbling up insects and caterpillars in the summer.

Shelter – many leaves and or stems shelter small creatures. Twiggy shrubs, wall-trained climbing shrubs, hedges, holes and ledges are ideal bird nesting sites. Dead tree trunks are like entire cities of small creatures. A pile of twigs and leaves makes a winter refuge for a hedgehog. A garden pond will house an entirely different community of animals.

Security – this basic need is often overlooked but is vital if you want the less common species in your garden. People and pets, or too much tidiness, will threaten the security of wildlife. Does your way of gardening offer security from chemicals?

WHAT YOU CAN GROW

British Trees – in order of size – oak, ash, alder, Scots pine, birch, cherry, rowan, crab apple. Choose according to the size and siting of your (and your neighbour's) garden.

Shrubs – hazel, hawthorn, holly, cotoneaster, berberis, buddleia, broom, lilac, (roses offer little sustenance, except to greenfly however ladybirds love greenfly!). Choose old-fashioned varieties or those nearest to the wild species.

Wall-Trained Plants – honeysuckle, ceanothus, wall-trained fruit trees. Be careful with creepers which can grow into problems, such as ivy or virginia creeper.

Herbaceous Plants – the choice is limitless! Aim to have something flowering and producing nectar throughout the summer and early autumn. Early flowers – crocus (birds eat them), primulae. Late flowers – lavender, heather, marjoram and other flowering herbs, daisy family (golden rod, asters, Michaelmass daisies, etc.), evening primrose. On the whole, old fashioned varieties make better wildlife plants than selected modern ones.

Ground Cover – creeping plants suppress weeds and keep the soil moist and comfortable for worms. Many varieties are available in garden centres. Try wild strawberries and share the fruit with the thrushes.

Grass and Lawns – a mixture of different grass species with clover (for bees) is best. Leave some patches of grass to grow long (cut with a scythe or sickle).

Wild Patch – could your garden spare a patch which can be left to grow nettles (for butterflies), where grasses and thistles can go to seed (for finches), where an old log can quietly moulder away (for fungi and hundreds of small creatures), where twigs and other nesting materials can accumulate . . . where stones are left unturned to gather moss?

Tree Care

It's a strange fact that when it comes to hair, many people would hesitate to get as little as a trim from any but the very best hairdressers. Yet badly cut hair is outgrown in a matter of weeks; badly cut trees bear the marks for ever. If you think a tree needs pruning or cutting, get advice from a qualified tree surgeon (look in Yellow Pages). Beware of casual loppers; trees deserve only the best!

There are just a few practical ideas to think about but for instructions for wildlife gardening, see book list on page 54.

Growing Vegetables Organically

Growing vegetables organically goes hand in hand with wildlife gardening. Organic gardening is not going back to quaint old ways. It is a system which combines tested and tried gardening tradition with scientific knowledge of the living processes in the soil. For detailed practical information you will need to consult books on organic gardening (see booklist page 54). Here we look at some basic ideas:

Caring for the soil structure – once the soil is free of 'big' weeds (docks, dandelions, couch grass, nettles, etc.) you should aim to dig as little as possible. Digging disturbs the soil structure built up by plants and soil organisms and brings buried weed seeds to the surface. Some organic gardeners never dig!

- *Avoid exposing bare soil to sun and rain for long periods.*
- *Avoid trampling and compacting. Walk along vegetable rows on old boards, or grass cuttings.*
- *Allow a different patch of the vegetable plot to rest each year, growing grass or 'green manure' (see below).*

Feeding soil and plants using less or no commercial fertiliser:

- *Make compost from weeds, leaves, kitchen waste. First read about composting! Alternatively, bury compostable waste in trenches (1-1½ ft deep). This is easier and quicker than making compost, particularly in winter.*
- *Use 'green manure'. This isn't an animal manure, as the name might suggest, but plants, chopped and dug straight into the soil – nettles, comfrey leaves or seaweed are classical green manure plants. Some organic gardeners grow crops directly for green manure.*
- *Use mulches. Mulching simply means covering soil, especially over the roots of plants. Mulches of compost, drying weeds, tea leaves or grass cuttings will add to soil fertility, encourage worms, hold in moisture and suppress weeds.*
- *If necessary, use bone meal or other commercially prepared organic fertiliser. Sprinkle wood (not coal) ashes for minerals.*
- *Make liquid plant food by soaking a lot of nettles in a little water (a week or longer). Dilute the nutritious, though smelly, brew and water straight on to plants as the nutrients are absorbed by the leaves.*

Keeping in soil moisture and managing with less or no watering:

- *Use mulches liberally, plant ground cover plants. Bare soil is unnatural.*
- *Store rainwater from the roof. It is better for plants than treated tap water and will help to conserve the public water supply in a dry period.*

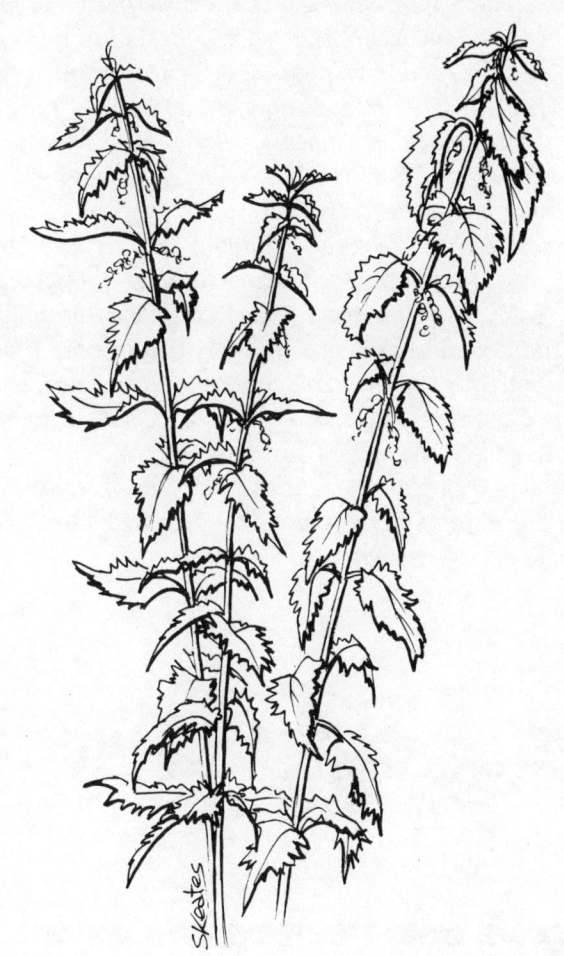

Managing weeds with less or no chemical herbicides:

- *A GOLDEN RULE is to prevent weeds from setting seed. Remember, 'one year's seeding, seven year's weeding'. Just chop off unripe seed heads if you can't keep up with the weeding. Minimum digging avoids bringing buried weed seeds to the surface.*
- *Keep soil mulched to suppress weed germination (see above). Thick layers of newspaper or a layer of black plastic are very efficient for this purpose. Chopped wood bark looks nicer but is more expensive. Hoe where it is impractical to mulch.*

Managing pests with less or no chemical pesticides: it may be difficult to manage entirely without chemicals to control pests, especially if the balance of nature outside your garden has been upset by other's use of chemicals. Try the following:

- *Mixed planting. Plant rows of vegetables and flowers alternately.*
- *Choose pest resistant varieties which are often the old-fashioned varieties.*

53

- *Avoid chemical fertilisers because plants grow excessively soft and succulent, with nitrate filled sap which pests love. Encourage pest predators – hedgehogs, frogs, toads, birds, bats, certain insects, ladybirds, wasps, hoverflies, to name but a few. This is where a wildlife garden scores!*
- *Protection devices – sunken jars of beer or upturned grapefruit skins to trap slugs; dry sand or ashes round seedlings to deter slugs and snails. Protect small plants and seedlings with rings cut from plastic bottles using serrated scissors. Great scope for improvisation here! (see YOUNG PEOPLE page 39).*
- *Use natural pesticides such as Derris or Pyrethrum in preference to chemicals, but use sparingly nevertheless.*

SOME USEFUL ODDS AND ENDS

- *Save holed rubber kitchen gloves to use in the garden. Buy cotton liners: these are cheapest from industrial clothing shops.*
- *Cut white plastic bottles into strips for durable labels, make holes with a leather punch. Write with a fade resistant pen.*
- *Collect sump oil in an oil drum and soak garden stakes over winter. Use sump oil to preserve timber in places where it won't be touched (it dries slowly).*
- *When on holiday by car in Europe, visit a gardeners/small holders store. you will find an extensive range of well-designed, robust garden equipment. A good continental scythe is one of the best tools for a large garden.*

The Garden Show

Does your local garden show provide for the organic grower to compete? Or do the prizes always go to the biggest vegetables, exhibition variety, chemically grown? Organically grown vegetables should be judged by standards appropriate to the aims of organic gardening, namely table quality. When you enter organic produce, make this clear to the judges and the public. Encourage the organising committee to promote showing organic produce in future garden shows.

ORGANIC GARDENING by Lawrence Hills, Penguin 1977

A MONTH TO MONTH GUIDE TO ORGANIC GARDENING by Lawrence Hills, Thorsons 1983

PLANNING THE ORGANIC VEGETABLE GARDEN by Dick Kitto, Thorsons

PLANNING THE ORGANIC FLOWER GARDEN by Sue Stickland, Thorsons

COMPOSTING: THE ORGANIC NATURE WAY by Dick Kitto, Thorsons

BACK GARDEN WILDLIFE SANCTUARY BOOK by Ron Wilson, Penguin Handbook

GETTING AROUND

... merrily we roll along ...

Independence is a cornerstone of our way of life and to be able to come and go as we please is high on our list of aspirations. The private car is a powerful means to that end; it is also a powerful means of wrecking the environment. Cars have already run up a big energy bill in manufacture, long before the first drop of fuel gushes into the tank. Once on the road, even the most economical car guzzles energy when measured against domestic fuel use; one litre of petrol keeps the average car going for 12 to 15 minutes but would heat a room for 4 to 5 hours. Motor vehicles pollute air (with acid and greenhouse gases, and hydrocarbons), the land (with dust, oil, some lead and rusting cars) and the sea through massive oil spillages from tankers. Road building changes landscapes and land-use and is fed by aggregate from gigantic, super-quarries, carving away the scenery in beautiful and environmentally sensitive areas of Britain. Add to this the services that motorised transport requires – service stations, traffic lights, road signs, policing – and on top of this the human cost of accidents. It is clear that the price of extreme independent mobility is phenomenal.

Before the invention of the combustion engine, transport didn't greatly affect the environment. Horses, donkeys and mules are fuelled by grass and oats (solar power), sailing ships are wind powered and don't pollute the sea, carts and carriages were basically constructed of wood (British hardwoods in those days) and little metal. Roads of beaten earth and cobbles, while they must have been horribly uncomfortable for travellers, weren't a hazard or barrier to wildlife and when abandoned would soon revert to usable land. In contrast, motorised wheels, fast and comfortable, require hard engineered surfaces which form permanent barriers, crossed at great risk by people and wildlife. Along these criss-crossing ribbons of tarmac, rolls a never-ending procession of noisy, dangerous and polluting machines, one every 10 metres if all 25 million vehicles in Britain were on the roads at one time.

Even though we know how much environmental damage cars (and all motorised transport) are answerable for, most of us feel that we cannot reduce the use of cars by very much. For such a step to become feasible, it would need an expansion of public transport as never before and a complete restructuring of our ways of living. It looks unlikely to happen until we are forced into it. But as the problems are growing all the time, we should each do what is possible meanwhile.

WHAT YOU CAN DO

Many effective ways to help require NO TECHNICAL KNOW-HOW whatsoever.

- *Minimise mileage!*

- *When you go for the papers or a loaf of bread don't take nearly a ton of metal with you, there and back! Short journeys with a cold engine are heavy on fuel.*

- *Shop locally so that you can go on foot or by bike. Locally grown food and locally made goods cut out long distance transportation by road.*

- *Use public transport as often as possible instead of driving. A sizeable reduction of pollution, acid rain and global warming would be made if cars were used only for leisure and other occasional purposes, instead of for daily commuting.*

- *If commuting to work by car is unavoidable, then club together with others and make a car pool. Please BE CARFUL!*

- *Cut your mileage by being well organised. Don't jump into the car to call round,*

TELEPHONE FIRST! Don't just chance it. If you have to drive across town, think of all the things that you could usefully do on the way and which could save you a journey on another day.

- *When you need to travel long distances, look first into the possibilities of going by rail; trains are the most energy efficient form of mechanised transport; plan ahead so that you can take advantage of economy rail tickets like Apex.*

- *Press your local authority for safe bicycle routes.*

- *Avoid all windscreen cleaners and de-icers in spraycans; a half potato works very well.*

- *Display some environmental stickers on your car.*

Here are some things to do that require A LITTLE TECHNICAL KNOW-HOW and DRIVING SKILL.

- *Drive with a fuel-saving, but safe, driving style – accelerate and brake gently, don't rev up the engine while standing still, cruise at 60, not 70 mph. Keep a running record of mileage and fuel consumption to check how well you and the car are doing. A drop in mpg can warn you of trouble with the engine.*

- *Choose your next car on the basis of fuel efficiency, less pollution, longer lasting engine and body, fewer extras (that can go wrong) and as always, safety features.*

- *Suggest to your company that they should run fuel efficient company cars which use unleaded petrol or diesel fuel.*

- *If you change oil yourself, pour the dirty oil into a tin and return to a garage. Never dump it anywhere or pour it down the drains. Old oil is useful as an outdoor wood preservative. Soak garden stakes in a drum of old oil.*

Finally some TECHNICAL possibilities.

- *Check the carburettor and ignition for efficient use of fuel.*

- *If you have not already done so, adjust your engine to using unleaded petrol. This is a practical proposition for the majority of cars.*

- *Consider fitting a catalytic converter to convert exhaust gases into less harmful form (this is law in some countries). Do this only if you regularly drive long distances.*

- *Rust-proof your car extremely thoroughly, especially the concealed parts of the body and keep every moving part well oiled.*

(The Green Consumer Guide by Elkington and Hailes has a very helpful section on the technical side of cars and conservation.)

And finally . . .

PS. *Tuck some polybags into a car pocket in case you unexpectedly need to buy things!*

PPS. *Never throw any rubbish out of the car window. Use one of the polybags for collecting rubbish.*

A DAY IN THE COUNTRY

Out in space, an extra-terrestrial being is looking out for signs of life on a small planet called Earth. Hovering over Europe or North America, IT will observe tides of shiny coloured capsules streaming into city centres as the city moves out of the earth's shadow into the light. Every seven days, this early morning tide is reversed and flows outwards from the cities into the nearby land and coast. Then, instead of the capsules ejecting one or two sombre coloured mites who scurry into buildings, these Sunday capsules empty out small

groups, livelier and more colourful, who move off into the surrounding countryside, often with small energetic specks dashing in circles round them. On some days, when visibility from space is particularly good, our extra-terrestrial being may count as many as 5 million of these capsules streaming into the country from the bigger cities of Britain. Just a fanciful Sci-Fi picture? Not at all. On fine Sundays or holidays the influx of cars, people and their dogs into the popular areas of countryside can amount to an invasion of wheels and feet.

Where many people want to enjoy and feel restored by the amenity of countryside, coast or mountains, there is always a dilemma – will these amenities be destroyed by our enjoyment of them, even just by our presence? Many feet will leave the ground eroded and trampled: picnickers may leave tell-tale traces of litter and burnt out campfires. People taking short cuts across farmland may damage crops.

These are examples of physical impacts on amenity; the impact on Nature is more insidious. Merely the presence of people, even quiet, considerate people, may disturb the lives and privacy of the shyer, more sensitive animals. Roads which provide access to the countryside cut across trails of deer, hedgehogs, frogs and toads and cause millions of animal road deaths. Recreational activities like hill walking and orienteering, skiing, mountain cycling can cause soil erosion, damage to plants and disturbance to animals. Wildlife deaths from eating lead shot or lead fishing weights are now less frequent, but are still recorded.

It is difficult to reconcile the best interests of people and wildlife; both sides will have to concede something but we can at least use our knowledge to be as helpful as possible to Nature.

WHAT YOU CAN DO

What you can do on an outing to the country is summed up neatly in a little code written by an unknown author (or poet!) some decades ago:

'**Take nothing but memories**' so, don't pick or dig up wild flowers (they wilt much faster than cultivated flowers); never take birds' eggs or nests; never collect butterflies or beetles. Little harm is done by picking blackberries, fungi or nuts in modest quantities provided no damage is done in collecting. Make sure you leave plenty of mushrooms and edible fungi growing to provide spores for next year's crop.

'**Leave nothing but footprints**' so, don't leave a scrap of litter behind, not a cigarette packet, not a film package, not a can ring – if you can carry things out full, you can carry them back empty! If you need to 'disappear behind a bush' bury all, including paper, or cover with a heap of leaves or grass.

'**Shoot nothing but pictures**' The only case for shooting that is compatible with conservation, is culling by wildlife rangers to keep the balance of numbers between species. This is, of course, only necessary because we, humankind, have upset Nature's balance.

'**Kill nothing but time**' so, for the same reason, don't kill any creature, by any method – trapping, snaring, ferreting, gassing, not even fishing, unless it is necessary for conservation.

Do you know that carving initials on a living tree trunk may be the starting point of fungus attack which will shorten the life of the tree and ruin the timber?

Some people think they are being helpful by destroying 'poisonous' fungi. This is not so. All fungi are a valuable part of the plant and animal community. Don't collect fungi to eat unless you are an expert.

This little countryside code has only four lines. Perhaps one could usefully add some more:

'Trample nothing but your toes' *meaning tread lightly on the plants that carpet the ground. Don't trample crops, ploughed fields. Stick to footpaths. Soil erosion and gullying is caused by excessive trampling.*

'Disturb nothing, not even your thoughts' *meaning don't disturb animals and birds, especially in the breeding season. Make as little noise as possible. YOU listen to the countryside instead of the countryside listening to YOU!*

'Chase nothing but your tail' *meaning keep your dog under control!*

'Burn nothing but enthusiasm' *meaning don't light any fires in dry weather, watch cigarette ends, don't leave glass around which can start a fire by focusing the sun's rays. If you do have a camp fire, under suitable conditions, make sure it is watched all the time.*

Keep the peace of the countryside and all who live there.

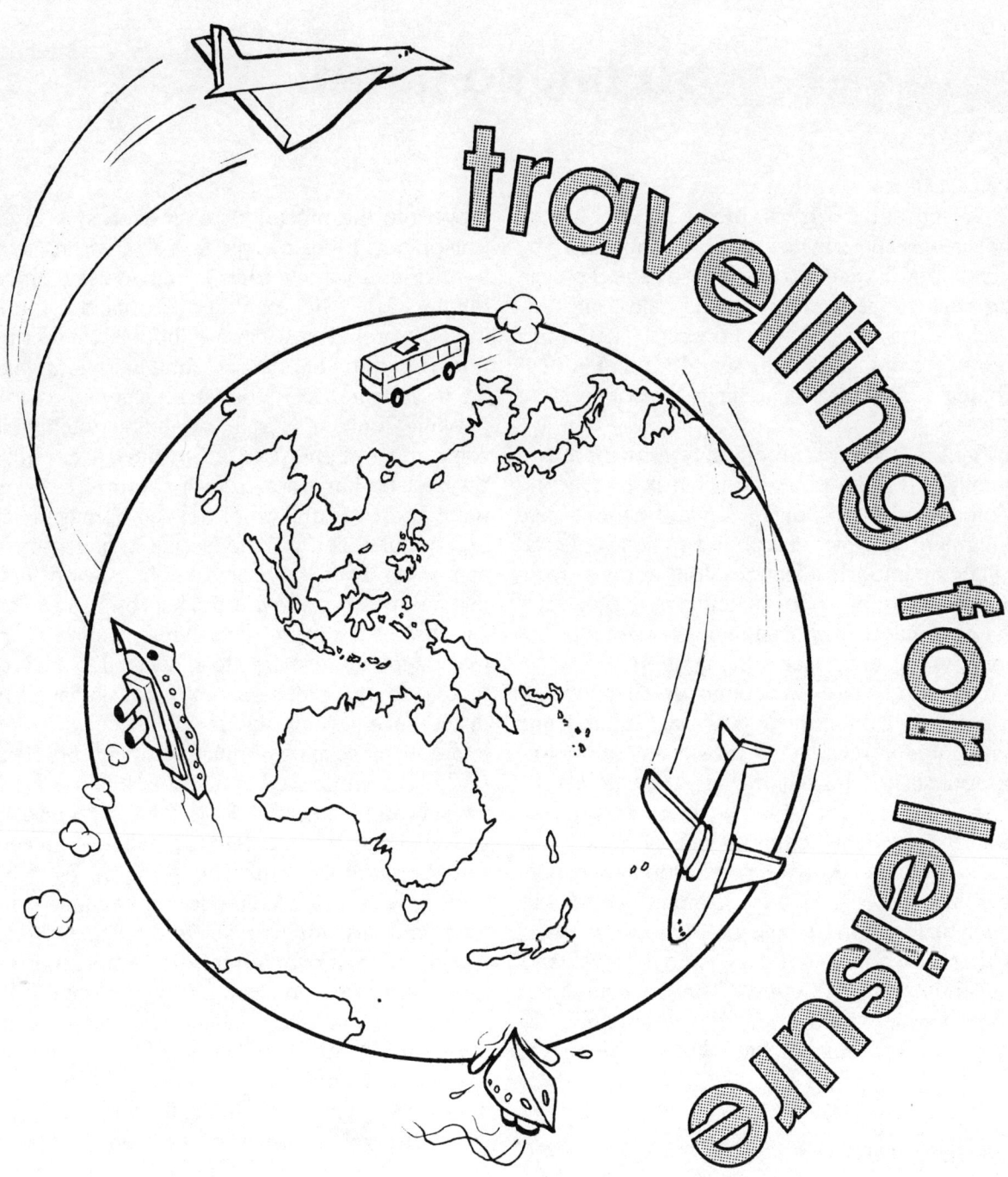

travelling for leisure

TRAVELLING FOR LEISURE

So far, so good!

Travelling and holidays abroad are among the most rewarding activities made available to us by technology. Experiences that used to exist only in the realm of saga and travellers' tales, are now easily accessible to many people. The tourist industry is opening up the world, so that 'globe-trotting' now admits to no limits of geography or climate.

Many of us have more leisure time than ever before and more of us are using it for tourism. More people are looking forward to an active retirement in which travel and holidays abroad will be an important feature. Visits abroad are an accepted part of school life and a period of travel between stages of education or between jobs has become a goal for many young people.

Tourism can be an economic benefit to the host country, but often at great social, cultural and environmental cost. While the sun shines brightly on tourists and holiday makers, they are casting lengthening shadows over the environment. Natural environments, often of highest quality, are being destroyed by rampant development of holiday facilities. There are examples worldwide. Over half of Spain's coastline is now exploited, with little concern for the ecological value of wetlands and sand dunes. The opportunity to earn money from tourism is tempting the Nepalese to abandon crop cultivation and to move into the precarious cash economy. In the Philippines, tribes-people find it more profitable to sell their spears to tourists than to use them for hunting fish. The employment that the tourist industry provides, welcome though it might be, undermines traditional skills and lifestyles, which are then degraded to tourist attractions.

While tourism is still expanding worldwide, some governments now see the need to control it. Switzerland has stopped further tourist development and construction of holiday homes in several popular valleys. The Nepalese government, as a measure to slow down deforestation, have taken steps to ensure that trekkers bring adequate supplies of kerosene with them. Tourists to the Galapagos are restricted to a few limited areas on the islands to minimise ecological damage and disturbance to their unique wildlife.

So, if mass tourism inflicts so many problems on host countries, what about travelling? Backpackers and trekkers prefer to think of themselves as travellers. What difference is there between travelling and tourism? The word 'traveller' implies a respect and feeling for other cultures and environments, and a sincere attempt to approach people on their own terms. Ever remoter parts of the world are being sought by the young, the active and the adventurous. One issue of the BBC's magazine *Wildlife* carried advertisements of twenty-two trekking companies offering safari-style holidays in Africa, India, the Far East, South America . . . in short, the globe! What impact on the people of remote areas will the presence of well-equipped, western travellers and trekkers have, even the most discreet? In photographing indigenous cultures, one may be documenting them but one is also contributing to their decline. We are faced with a real dilemma, since travel is one of the most effective ways of educating people and building a solid basis for peace between nations.

Tourists to developing countries can be targets for souvenir merchants offering products from animals on the CITES list of endangered species (see page 73). Ivory objects may be from legal sources but could be poached; could you be sure? Crocodile leather articles could come from a crocodile farm (legal, but inhumane) but may equally come from wild crocodiles; could you tell the difference? Shells and corals may still be plentiful in some places but will become rare if many are collected as souvenirs.

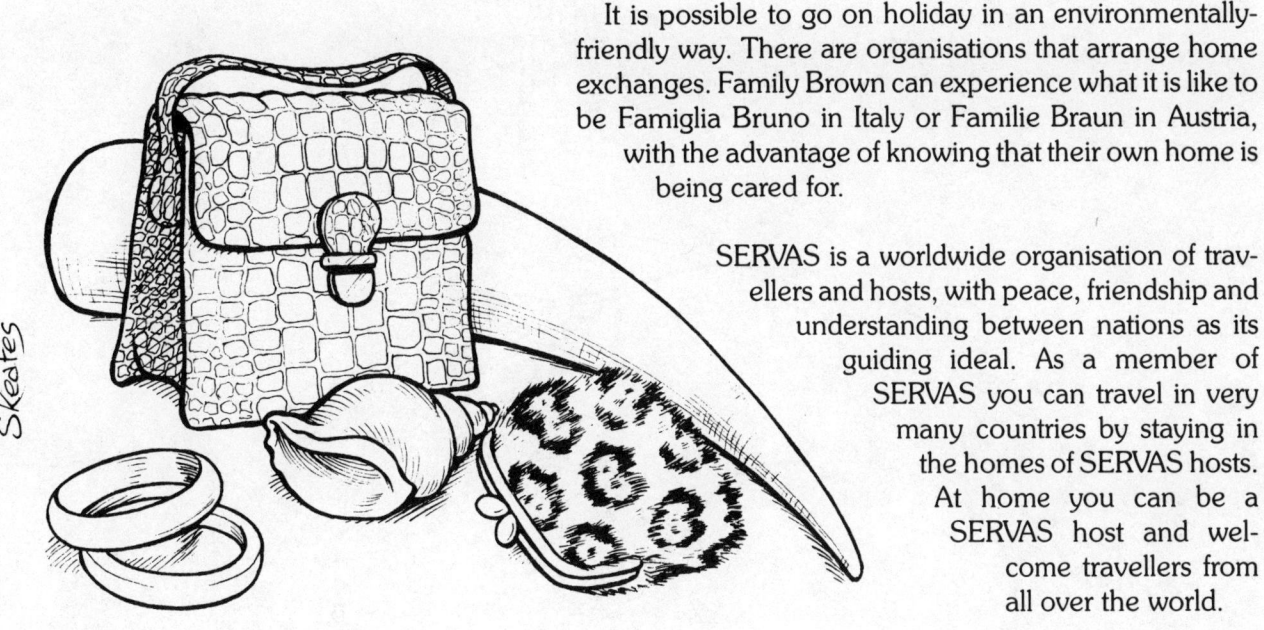

It is possible to go on holiday in an environmentally-friendly way. There are organisations that arrange home exchanges. Family Brown can experience what it is like to be Famiglia Bruno in Italy or Familie Braun in Austria, with the advantage of knowing that their own home is being cared for.

SERVAS is a worldwide organisation of travellers and hosts, with peace, friendship and understanding between nations as its guiding ideal. As a member of SERVAS you can travel in very many countries by staying in the homes of SERVAS hosts. At home you can be a SERVAS host and welcome travellers from all over the world.

WHAT YOU CAN DO

- *Travel on holiday by train or boat where possible; use public transport while on holiday and enjoy the company of local people. Air travel is the most fuel consuming form of transport and puts fuel combustion products directly into the sensitive upper atmosphere. For most overseas and all long distance travel, flying is regrettably usually the only practical option.*
- *Try to minimise your impact on the environment and local people wherever you go. Keep a 'low profile', (see also A Day in the Country, page 59).*
- *Do nothing, buy nothing, that will further endanger wildlife. If in doubt about a souvenir, don't buy it; if in doubt about a sport, find some alternative.*
- *If you buy souvenirs while visiting a developing country, buy if possible from recognised craftsmen or craft co-operatives such as those supported by churches or agencies such as Oxfam, Traidcraft, CAFOD, Action Aid and Christian Aid.*
- *Find out about house exchange organisations.*
- *Write to SERVAS for information about becoming a traveller or host.*

Books for further holiday ideas:

HOLIDAYS THAT DON'T COST THE EARTH, Julia Hailes and John Elkington.

THE GOOD TOURIST, Syd House and Katie Woods.

The National Secretary
SERVAS, England, Scotland & Wales
PO Box 1035
Edinburgh EH3 9JQ

65

investing

for the
environment

INVESTING FOR THE ENVIRONMENT

One of the most potent forces shaping the environment and one that causes much exploitation of people and resources, is investors' expectations of maximum return on their investments. All companies operate on the principle that 'our first duty is to MAXIMISE PROFITS for our shareholders'. So if running a business with concern for the environment and people means less profits, few companies would seriously consider doing so, even though survival into the future of all commercial enterprises depends on conservation now. Businesses are too often forced to ignore the long term in order to stay competitive in the short term.

These are the bare bones of the situation which faces investors and companies who would like to get off the environmentally and socially destructive treadmill. The opportunities for ETHICAL INVESTMENT meets this need.

Investing ethically means making your money work to promote projects with a positive social value and a low impact on the environment. It might mean investing in the development of environmentally helpful technologies.

It could mean avoiding companies with a poor record for pollution, or exploitation in the developing world; or boycotting any businesses connected with armaments or nuclear power. A further step would be to buy a share in Traidcraft (address page 80). Or to invest in Shared Interest, the UK's first 'social investment society' (address page 80) and thereby invest in sustainable jobs for people in need, especially in the developing world.

Most ethical funds have done as well as any other funds, with the exception of those like Mercury Provident which specifically seek to provide low-interest loans.

It is best to invest through an ethical unit trust because they can do the necessary research to find out about companies' practices. However the financial complexities of the business world make it difficult even for ethical unit trusts to completely disentangle their investments from areas that their clients may wish to avoid. EIRIS, Ethical Investment Research Service, has produced a guide, 'Choosing an Ethical Fund'. There are now several advisory services which can help with ethical investment.

WHAT YOU CAN DO

- *Consider putting at least some of your savings into ethical funds and know that you are helping socially useful and environmentally sound businesses to flourish.*

- *When starting a pension/insurance scheme, consider doing this through an ethical fund.*

- *Ask you investment consultant about ethical investment. Every enquiry helps to give it publicity.*

- *Consult WHICH? February 1989 'A Fair Profit?'*

- *For general advice on ethical investment write to*

The UK Social Investment Forum
Keeley House
22 - 30 Keeley Road
Croydon CR0 1TE

- *Write to EIRIS, The Ethical Investment Research Service for their guide, 'Choosing an Ethical Fund'.*

EIRIS FREEPOST
LONDON
SW8 1BR

- *Is there a community enterprise in your area which you could support, e.g. a village of community workshop, capitalised by its users, suppliers, neighbours, workers, etc. This would be a very green and effective form of investment.*

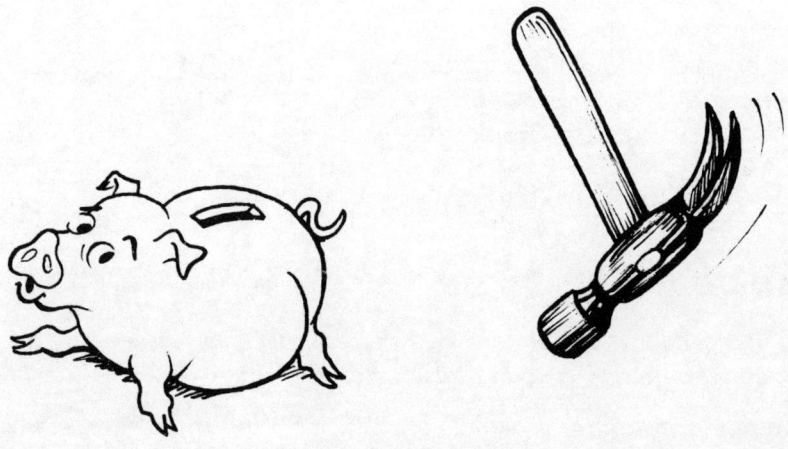

happy birthday

MERRY CHRISTMAS

HAPPY BIRTHDAY!
MERRY CHRISTMAS!

Choosing presents is a delight – or a headache! Whichever it is, a lot of money is spent on presents for Christmas, weddings, christenings, home-comings, visits, parties . . . has anyone added up what is spent on birthdays alone each year, or what amount of resources the gift industry uses up?

It's been customary for many years to buy Christmas cards in aid of good causes. Charities raise a lot of funds this way. You could go one step further and also buy presents to support socially and environmentally useful organisations.

If you have never yet visited a 'developing world Fairtrading' shop (what a mouthful! but 'charity shop' is surely an obsolete name? You have a treat in store. You will find an irresistible variety of hand-made goods, such as are found nowhere in other shops, all very reasonably priced. There are both useful and decorative handicrafts, jewellery, furnishing items, clothes, all made in small workshops in developing countries, using every available traditional material –

silk, jute, cotton, leather, wood, silver, ceramics, traditional metals. . . . These workshops are supported by Fairtrading organisations and are a lifeline for hundreds of rural communities in developing countries. Before going shopping elsewhere, look into one of these shops.

TRAIDCRAFT and OXFAM are the best known 'fair-trading' organisations (see page 80). Traidcraft sells mainly though mail order and a network of local representatives. Although Traidcraft have only a few direct outlets, many groups sell their goods through local organisations such as church groups. Oxfam now has shops in most areas. Other environmental groups also sell Fairtrade goods by mail order (Greenpeace, Friends of the

Earth, Action Aid, Amnesty International and many others).

There are now green mail order businesses, with catalogues of goods that are 'good for the individual, good for our community and good for our plant', to quote from one of them. Among the many practical items are many which would make good presents (see page 80).

Meanwhile, we mustn't forget that goods are also hand-made in Britain. By buying from local crafts people you support local talent and small businesses and help us to move in the direction of quality, self-reliance and a restored environment.

WHAT YOU CAN GIVE

Here are some suggestions for further environmentally helpful presents:

- *A great big shopping bag or basket, to hasten the retirement of the polythene bag.*

- *Tools, to encourage making and mending things.*

- *Rechargeable batteries and battery charger, to reduce wastage of valuable metals and to avoid the hazardous pollution they cause in waste dumps. You can recharge these batteries a 1000 times.*

- *House and garden plants, anything from a small tree (native, of course!) to herbs growing in a pot.*

- *A bird feeder (with food containing grains, not peanuts).*

- *Recycled paper goods. Look out for an eco-label.*

- *Cosmetics and toiletries from plant products.*

- *A subscription to an environmental organisation (see page 77).*

- *A subscription to an environmental magazine (see page 80).*

- *One of the many stunning environmental picture books now available in all bookshops.*

- *A bag of Brazil nuts, to support good rainforest products.*

- *Organically grown fruit, preferably British.*

- *Always take a bottle of organic wine to a party, or non-alcoholic Norfolk Punch, made with dozens of different aromatic herbs.*

Many Happy Returns Planet Earth !

MEANINGS OF ENVIRONMENTAL TERMS

The following words are frequently used in writing or talking about the environment. Many have been used in this book but some others have also been included.

ACID RAIN: oxides of sulphur and nitrogen react in the atmosphere to form an acid pollution which come down either dry or as acid rain. This pollution damages trees, acidifies lakes and kills fish and corrodes away old stone buildings and sculptures. There are many sources of sulphur and nitrogen pollutants – vehicle exhausts, burning of fossil fuels in homes and power stations, ore smelters, intensive agriculture.

ADDITIVES are chemicals used in the processing of food and drink, as preservatives, to lengthen shelf-life, for flavouring, to fortify (added vitamins, trace elements) and for cosmetic reasons. We give symbolical colours to foods; butter and margarine must be bright yellow, strawberry jam or yoghurt bright pink, raw meat bright red, white bread really white. Natural foods are rarely such strong colours, so as we begin to value additive-free foods, we must adjust our expectations.

AGRIBUSINESS is farming conducted on a large scale where maximum efficiency and competitiveness are the first goals. Cultivating huge areas of single crops and raising animals by factory farming are features of agribusiness. Agribusiness depends on using AGROCHEMICALS and AGROTECHNOLOGY and valuing the soil and farm animals only as efficient producers of food.

AGRO-CHEMICALS are, as the word suggests, the chemical compounds that are used in huge quantities in most agriculture – compounds of potassium, nitrogen and phosphorus for fertiliser, pesticides to kill animal pests, herbicides to kill plant pests (weeds!), fungicides to kill moulds. Well over 2 million tons of chemical fertiliser are used each year – at the same time that sewage and farm slurry is being expensively disposed of.

ALUMINIUM is the most abundant metal on the planet but extraction from the ore (bauxite) needs large amounts of electricity and is very polluting. Recycling uses 20 times less energy than producing new aluminium.

APPROPRIATE TECHNOLOGY is technology designed to be used and maintained by people on a human scale. It aims to use RENEWABLE energy where possible but can also include the use of small scale, advanced equipment where the need arises.

BIODEGRADABLE: substances are biodegradable if they can be broken down by micro-organisms. They are DEGRADABLE if they break down through exposure to light and water. Both processes cause materials to disappear but sometimes the products of degradation can be polluting. Plastic litter can be picked up; harmful degradation products cannot.

BIODIVERSITY (GENETIC DIVERSITY): the astronomical number of different life-forms and their infinite range of size, shape, colour and activity is the most valuable asset of this planet. This diversity is necessary to keep conditions on earth suitable for life.

CASH CROPS are crops grown primarily for selling. Selling one's surplus has always been a basis for local trading but growing crops for distant markets, as is done on a big scale in the developing countries, becomes a burden rather than an asset. The best land is given to cash crops, leaving farmers to feed themselves from poor, marginal land. Often the crops grown are foreign to the area and won't grow without a lot of chemicals, forcing cultivators on to the pesticide treadmill. The world price for most cash crops has been falling for years and will be forced down further under the new GATT agreements.

BIOSPHERE: the living world with the physical structures (water, air, soil, rocks) which support it.

CONSERVATION means using and managing resources in a way that they remain available to foreseen generations. Conservation means conducting human life in such a way that all species of animals and plants can exist in sufficient numbers to breed. In conservation, it is of great importance to preserve HABITATS.

C.I.T.E.S: Convention on International Trade in Endangered Species. Countries who have signed this agreement prohibit trade in endangered species.

CONSUMER: The recent change from using the time-honoured word 'customer' to the word 'consumer', to describe the buyer of goods and services, could say a lot about attitudes. It is not very complimentary, suggesting an unsatiable devouring of goods.

DEFORESTATION and AFFORESTATION: deforestation is the removal of natural forests, through big scale logging operation, fire or continuous, piece-meal clearance for agriculture. In the tropics, deforestation mostly leads to DESERTIFICATION once the fragile soils are exposed to the powerful tropical sun and rain. Afforestation usually means planting MONOCROPS of timber or pulp trees in plantations, not the restoration of natural forest.

DESERTIFICATION is the end-result of exploiting the land in dry, tropical parts of the world, usually the result of damaging the tree or grass cover to a point where it cannot regenerate. Nearer to home, the Scottish Highlands (but also much of Wales and upland England) have been called a 'wet desert' because the prevailing land-use does not permit the regeneration of the natural forest cover.

DEVELOPMENT: the dictionary defines it as 'process of growing and progressing'. It is an ambiguous and confusing word because different groups give it a different meaning and have opposing ideas on the aims of 'growing and progressing'. When government agencies talk about development, they mean economic growth and industrial expansion. When the World Bank talks about 'World Development' it means mega-technological projects such as big dams. Organisations like Oxfam, Action Aid, Christian Aid, CAFOD, use the same word, development, for their work to promote self-help through APPROPRIATE TECHNOLOGY in the third world. They even call themselves 'Third World Development Organisations', although they have quite different aims and methods and their work is necessary, in large part, to mitigate the often disastrous consequences of government's form of 'development'.

DEVELOPING WORLD: the world can be, roughly, divided into the well-off one third of countries (Europe, USA, Australasia, Japan and possibly the CIS) and the badly-off two-thirds of countries (all the rest). We do not have an appropriate name for the two-thirds world. 'THIRD WORLD' is derisory and 'DEVELOPING WORLD', though preferable, begs the question of what sort of development they are following. Much of our DEVELOPMENT has been detrimental and our world has a lot of developing to do in better directions.

ECOLOGY is the study of the balance of nature; how plants and animals interact, depend on and compete with each other. The study of how humans interact with nature, and how their activities affect and are affected by nature, is called human ecology. An activity or life-style can be considered 'ecological', if it disturbs the balance of nature as little as possible.

ECOSYSTEM: a group of plants and animals that interact and support each other, together with the physical features of the environment that they depend on.

ENERGY EFFICIENCY: high efficiency in use of energy, especially from fossil fuels, is one of the most effective ways of combatting GLOBAL WARMING and many other forms of POLLUTION. We could substantially reduce fuel consumption without any fall in living standards.

ENVIRONMENT: a clumsy word for what surrounds us and supports our existence! The environment is not only outside us – we and our surroundings together are 'the environment'.

EROSION: the wearing away of soil and rocks by water, wind, ice and sun. Erosion is an infinitely slow natural process, crumbling mountains and enabling new rock strata to form under the oceans. DEFORESTATION dramatically speeds up erosion, particularly of soils. This is severest where land is mountainous and deforested as in Nepal, but in Europe soils are also eroding, at a rate of one billion tonnes each year (about 4 tonnes each).

EXTINCTION is a frightening word – the vanishing of a species from the face of the earth. There are millions of species but every extinction breaks a thread in the web of life. Humankind is maybe causing the extinction of as many as 10 species each day.

FAIRTRADING is the guiding principle of a number of ALTERNATIVE TRADE ORGANISATION (ATO'S) who trade with producer groups in the developing world. ATO's will pay a fair price for goods and will trade only with enterprises guaranteeing acceptable wages and working conditions for their employees. Traidcraft, Oxfam, Cafod and Twin Trading are the main ATO's in Britain. ATO's make ethical choice in a range of goods a reality for the consumer.

FISH FARMING is an age old tradition; in the Middle Ages villages had fish ponds and the fish mopped up village refuse. Today fish farming is a highly scientific industry and the growing number and size of farms threatens to become another ecological problem. Farmed fish are fed on meal made out of 'inferior' fish (inferior from a culinary, not ecological, point of view). The cages release pollutants into clear water, from excess food, excreta and potent pharmaceuticals needed to control disease among crowded fish. However fish farming could help marine conservation if managed by environmentally sound methods.

GLOBAL WARMING caused by the accumulation of GREENHOUSE gases in the atmosphere was predicted a hundred years ago by a famous Swedish chemist! The problem gases are carbon dioxide from burning fossil fuels and burning forests, methane from agriculture and waste dumps and the notorious CFC's used by industry and in aerosols. The quantities annually released into the atmosphere by these activities are phenomenal; 580,000 tonnes of CFC's, 270,000,000 tonnes of methane and a staggering 22,000,000,000 tonnes of CO_2. It hardly seems necessary to do research to ascertain that we are unbalancing the earth's atmosphere, and that this will have consequences, even if we can't predict exactly what they will be.

HABITAT: a special bit of the environment which has the right combination of features to sustain the life of a plant or animal species. A sensitive species can be destroyed by degrading just one feature of its habitat. See CONSERVATION.

HERBICIDE: chemicals used to kill weeds or to prevent weed seeds from germinating.

HUMUS is the organic part of the soil, formed by the breakdown of living matter. Humus gives soil its structure and fertility. it takes a very long time, sometimes hundreds of years, for humus rich soils to accumulate. The humus content is easily depleted by poor farming.

INTENSIVE FARMING is the practice of raising animals and crops for the market in the shortest possible time and with maximum economic efficiency. For animals this is usually achieved by FACTORY FARMING and for crops by high inputs of AGROCHEMICALS. EXTENSIVE FARMING, in contrast, reduces the inputs and may use no chemicals at all (ORGANIC FARMING). The farmer accepts lower yields, but has fewer expenses, and keeps within the natural constraints of the environment.

JUNK FOOD: fizzy drinks, crisps, sweets, chocolate wafers, most sausages are examples of foods that are better described as edible commercially manufactured items. They usually contain a lot of sugar, or salt, and a variety of chemical additives to lessen their cost and to make them appealing to eye and palate. Many children have an intolerance to these additives.

LUXURY: a word which suggests extravagance and a measure of indulgence! A product or an activity that makes a heavy demand on the environment is a luxury; all disposables are a luxury. So is flushing out our sewage to sea, or wrapping up small amounts of watery drinks in aluminium. Furniture, built out of the best materials to last for generations, is not a luxury, whereas cheap furniture that soon wears out is, from an environmental point of view, an extravagance.

MONOCROP: a single species planted over a big area. Monocrops of Sitka spruce cover big areas of hill country; a barley field is a monocrop. When pests attack, they can spread through monocrops at great speed, going straight from one plant (or tree) to the next unchecked. Mixed cropping slows down the spread of pests and diseases.

NATURE is a word that should really be reserved for those plant and animal communities that exist untouched by human activities. True wild nature scarcely exists anywhere now because few parts of the planet remain uninfluenced by our activities; not even the Poles! It is a profound social and ecological question, just how much of the planet should be left for nature and how much humans have a right to use.

ORGANIC FARMING: many people think organic agriculture is only farming without chemicals, but this is too narrow a view. Organic agriculture uses the natural processes of growth, decay and living interactions and is in particular concerned with building up the living fertility of the soil.

OZONE is a form of oxygen (O_2) in which three atoms of oxygen join together (O_3). Only a minute proportion of oxygen is normally in the form of ozone. Ozone is an asset, or a liability, depending on where in the atmosphere it is. It is undesirable in the lower atmosphere as it reacts with other pollutants to form smog; in the upper atmosphere it is essential to protect life from ultraviolet radiation. CFC's from industry or as aerosol propellants break down ozone into oxygen in the upper atmosphere, especially in the polar regions, causing areas of ozone depletion, ozone holes. CFC molecules can persist in the atmosphere for around 100 years.

PESTS AND WEEDS are species that can multiply very rapidly when we give them the opportunity, for example by upsetting an environment in which they are normally in balance with other species.

PHOSPHATES and BORATES are chemical compounds added to detergents to strengthen their cleaning power. They also strengthen their polluting power and are not used in environment-friendly washing products.

POLLUTION: to pollute means to load air, land or water with products, beyond the point at which natural processes can deal with them. Usually these are waste products but oil tanker spillages cause severe pollution with costly useful materials.

PRE- and POST-CONSUMER WASTE: the recycling industry makes a distinction between these wastes. Pre-consumer are wastes (off-cuts) collected in the manufacture of items like paper, plastic and metal goods. These are put back into the manufacturing process. Post-consumer wastes are what the consumer discards after use and therefore give genuinely recycled products.

PROCESSED FOOD: all cooking and food preparation which changes the form of the food, is processing but the word is usually applied to industrially prepared food. CONVENIENCE FOOD is mostly processed and contains additives to enhance taste, texture, appearance and keeping qualities. There is no reason why convenience foods shouldn't be of high nutritional quality and culinary standard.

QUALITY OF LIFE is a phrase used to describe the unmeasurable things in life, without which no STANDARD OF LIVING, however high, has any real worth. Things such as satisfaction, personal fulfilment, interests, skills as well as health, freedom from want and beauty. Standard of living refers to income, purchasing power, material consumption and other measurable quantities that can be expressed as statistics.

RENEWABLE ENERGY is the energy of the sun, the energy of wind, water and tides and the heat energy from the centre of the earth (geothermal). Using renewable energy does no harm to the environment (big dams excepted) as this energy flows continually anyway; using it merely holds up the flow temporarily. RENEWABLE RESOURCES are those materials derived from animals and plants, soil and water. They are available only if we use them sustainably, in such a way that they can replenish themselves.

SOFT WOODS, HARD WOODS: nature's division of the tree kingdom into coniferous (cone bearing) and broad leaf trees nearly matches our division of timber into soft and hard woods.

SUSTAINABLE DEVELOPMENT: here is what the Brundtland Commission (1987) says – 'development that meets the needs of the present without compromising the ability of future generations to meet their own needs'. Much depends on what is meant by DEVELOPMENT. If it is material economic development, which depends on economic growth, then it cannot be sustainable. If the phrase is used in the sense of small-scale, environmentally-considerate development, then it could be sustainable, provided that world population stabilises.

VEGETARIAN or VEGAN: everyone knows that vegetarians don't eat meat, that some don't eat fish and that most eat dairy products and eggs. Some people are vegetarians because they don't like meat but some because they object to killing animals. These vegetarians should remember that having dairy products and eggs depends on the slaughter of animals – what happens to all the male animals that are born? Vegans do not eat any food of animal origin and avoid wearing leather. Vegans should consider the need for (some) animals in farming, especially ORGANIC FARMING, where it is not certain whether soil fertility can be maintained indefinitely without animal manure.

WHOLEFOOD is food from which nothing edible and nutritious has been removed. Many wholefoods sold in wholefood shops are grown ORGANICALLY.

YOU and ME: the people who can make the difference!

SOME ENVIRONMENTAL ORGANISATIONS

Many of these organisations have been mentioned in this book. Some more have been added for further interest.

The following are national (and international) Environmental Organisations which you can join, to give their work your support and to receive information:

FRIENDS OF THE EARTH,
26-28 Underwood Street, London N1 7JQ. 071-490 1555

FOE is one of the biggest and longest established environmental campaigning organisations. FOE campaigns on many big global issues and also on all issues of local environmental concern. FOE (UK) supports local groups throughout Britain, FOE (Scotland) supports local groups in Scotland, FOE (International) co-ordinates groups and their campaigns the world over.

GREENPEACE UK, 30-31 Islington Green, London N1 8XE. 071-354 5100

Greenpeace is an international environmental organisation best known for its intrepid actions to stop whaling and nuclear dumping at sea. Campaigns to conserve marine life, to combat marine pollution, atmospheric pollution, nuclear power, trade in endangered species, are among its many activities. There are many Greenpeace support groups throughout the UK.

WORLD WIDE FUND FOR NATURE,
Panda House, Weyside Park, Godalming, Surrey GU7 1XR

Tropical forest conservation is one of WWF's most important areas of action. WWF is much involved with environmental education and publishes a lot of material for use in schools. It supports scientific research into conservation of biodiversity.

ECOROPA, Crickhowell, Powys, Wales NP8 1TA

Ecoropa is a Europe-wide campaigning organisation with an active branch in Britain. It publishes concise and informative leaflets on a range of issues such as defence and war, nuclear power, food and health.

SCOTTISH CAMPAIGN TO RESIST THE ATOMIC MENACE (SCRAM),
11 Forth Street, Edinburgh EH1 3LE

Relevant to the whole of Britain, this group does much more than its name suggests. SCRAM is now one of the leading proponents for appropriate energy sources and energy conservation. Newsletters on all energy matters, each one with a sharp and lively column by Little Black Rabbit.

UNITED NATIONS ASSOCIATION OF GT. BRITAIN AND N. IRELAND (ENVIRONMENTAL SECTION),
3 Whitehall Court, London SW1A 2EL

The UNA exists 'to promote the principles of the UN Charter and the role of the UN in international affairs'. The UNA also takes an active part in environmental conservation. It organised preparatory meetings in 20 cities, for the Earth Summit 1992, and follow-up meetings afterwards.

WOMEN'S ENVIRONMENTAL NETWORK
287 City Road, Islington, London EC1V 1LA. 071-490 2511

WEN has 'the aim of educating, informing and empowering women who care about the environment. They aim to provide a forum where women can speak out, with women's perspective, on environmental issues which affect us all.' WEN runs workshops, public meetings, and campaigns on environmental issues specifically affecting women. Local groups are active in several areas of Britain.

FUTURE IN OUR HANDS, 120 York Road, Swindon, Wilts. SN1 2JP

FIOH (UK) says: 'We are part of a popular international movement, founded in Norway in 1974, which recognises that all human beings are individually and jointly responsible for caring for the earth and sharing its resources. We are ordinary people who are pooling our ideas, skills, resources and goodwill in order to promote co-operation rather than competition between people and nations.' FIOH gives encouragement to people who are making a personal commitment to live a simpler life-style.

LIFE STYLE MOVEMENT, Manor Farm, Little Gidding, Cambridgeshire PE17 5RJ

'The Life Style movement exists to inform, encourage and support people who, as a sign of their commitment to global justice, try to live a more responsible lifestyle, to help bring about a just and ecologically sustainable society.' They write a four monthly newsletter.

HENRY DOUBLEDAY RESEARCH ASSOCIATION (HDRA), National Centre for Organic Gardening, Ryton-on-Dunsmore, Coventry CV8 3LG. 0203 303517

HDRA is an association for organic, and would-be organic, gardeners and growers. They send a quarterly newsletter, full of invaluable information on organic gardening. The centre at Ryton has a large demonstration garden, an attractive tea room and a shop selling books and gardening materials (also by mail order).

THE SOIL ASSOCIATION, 86 Colston Street, Bristol BS1 5BB. 0272 290661

This is a national association for the promotion of organic agriculture (HDRA works mainly with smaller growers). The SA defines standards for organic production and awards the Soil Association Symbol for foods produced to their standard. It acts as a consumer watchdog on food quality issues. The Living Earth is the membership organisation of the SA and produces a journal of the same name.

ROYAL SOCIETY FOR THE PROTECTION OF BIRDS, The Lodge, Sandy, Bedfordshire SG19 2DL. 0767 680551

The RSPB has a big and enthusiastic membership of all ages. Its aim is simply stated: 'the conservation of birds and their habitats'.

ENVIRONMENTAL TRANSPORT ASSOCIATION LTD, 15a George Street, Croydon CR0 1LA

ETA campaigns both on behalf of the environment and on behalf of the traveller! They want to see priority given to walking, cycling, trains and buses, more goods going by rail and water, reduction of exhaust emissions. ETA Ltd offers a number of services which include insurance and a Helpline if you run into trouble on any non-commuting journey. They issue 'Going Green', a bi-monthly magazine.

SUSTRANS, 35 King Street, Bristol BS1 4DZ. 0272 268893

Sustrans is short for Sustainable Transport. They aim to 'provide, demonstrate and promote environmentally sustainable alternatives to motor transport'. They build safe and attractive routes for pedestrians, cyclists and those in pushchairs and wheelchairs.

NEW CONSUMER, 52 Elswick Road, Newcastle Upon Tyne, NE4 6JH
ETHICAL CONSUMER, 16 Nicholas Street, Manchester, M1 4EJ

Ethical Consumer and New Consumer are separate but similar, not-for-profit organisations who research the social, ethical and environmental practices of many of the companies who produce our everyday products. Ethical Consumer publishes a bimonthly magazine; New Consumer issues a newsletter. Both organisations are dedicated to enlisting consumer spending power for the benefit of people and the environment. Both publications are packed with information to help you make shopping choices which are more environmentally and socially supportive. While the consumer information produced by each covers much the same ground, the organisations complement each other in the way they gather information. NC sends questionnaires to companies directly; EC collates information from a wide variety of already published sources.

The following are some environmental magazines not associated with organisations:

THE ECOLOGIST, Red Computing, 29A High Street, New Malden, Surrey, KT3 4BY (subscriptions)

The Ecologist is a bi-monthly journal (now at volume 22) which publishes articles by leading authorities. It brings new ecological problems to light and analyses their causes on sound, scientific lines. It is a solid, but always readable, reference journal on global, ecological issues.

NEW INTERNATIONALIST, 120-126 Lavender Avenue, Mitcham, Surrey CR4 3HP

On the cover, is printed 'The people, the ideas, the action in the fight for world development.' The NI reports uncompromisingly on issues of world poverty and inequality and focuses attention on the unjust relationship between rich and poor nations. And, it is an informed source of information on environmental issues. It has reached well over 200 numbers.

RESURGENCE, Salem Cottage, Trelill, Bodmin, Cornwall PL30 3HZ

Resurgence, now past its 150th issue, is a forum for writers on both ecological/environmental matters and on spiritual/alternative visions for the future. Not only ecologists, but also philosophers, artists, poets, and H.R.H. Prince Charles, are among its many contributors.

GREEN TEACHER, Machynlleth, Powys, Wales SY20 8DN

Green Teacher is a magazine which presents new approaches and thinking in environmental education to teachers and curriculum developers. It aims to develop new models of learning, with lots of lively practical ideas.

PRACTICAL ALTERNATIVES, Victoria House, Bridge Street, Rhayader, Powys LD6 5AG

'A magazine of ways of saving resources in everyday life, and a newsletter of Ecological Life Style Ltd.' And a very lively magazine too, with information on many practical aspects from combatting damp to combatting unfairness in our voting system, and from constructing an eco-loo to constructing an entire eco-village.

The following are some organisations which sell Fairtrade and other environmentally and socially useful products.

TRAIDCRAFT PLC, Kingsway, Gateshead, Tyne and Wear NE11 0NE

Traidcraft's motto is 'Trading for a Fairer World'. They are one of the Alternative Trading Organisations (ATO) whose aim is trade based on 'fair share, concern for people and care for the environment', thereby promoting more equitable trading relations than prevail today. Traidcraft's range includes household items, crafts, jewellery, clothes, toys, paper products, foods and many, many other most attractive items. They sell through more than 100 Traidcraft shops and 1,400 voluntary representatives throughout Britain, or you can write for a mail-order catalogue.

SHARED INTEREST, Freepost, Newcastle Upon Tyne, NE4 5BR

SI describes itself as a Friendly Society and offers a 'unique channel through which people in the UK can invest in sustainable jobs for people in need, especially in the Developing World.'

OXFAM, (Head Office), Oxfam House, 274 Banbury Road, Oxford OX7 7DZ. 0865 56777

Perhaps the best known 'Third World Development Organisation', Oxfam has shops in most towns and cities. They welcome gifts of saleable clothes, books, records and bric-a-brac, whose sale supports Oxfam's relief and development work. Oxfam is also an ATO and many of the shops carry a range of crafts made in workshops in the developing world and in sheltered workshops in Britain. Oxfam publishes information about the developing world.

THE GREEN CATALOGUE, Freepost (BS 7348), Axbridge, Somerset BS26 2BR. Tel: 0934 732469. Fax: 0934 732748

This is a commercial enterprise which has the aim 'to help you cut through all the "environmentally-friendly" claims and choose green products that you need for your life.' The catalogue has a big range of products from green nappies, low energy light bulbs, unbleached cotton items, rainforest products, recycled stationery, to accessories and gift ideas. By mail order only.